Environmental Education, Ethics and Action in Southern Africa

Environmental Education, Ethics and Action in Southern Africa

HUMAN SCIENCES
RESEARCH COUNCIL

ENVIRONMENTAL EDUCATION
ASSOCIATION OF SOUTHERN AFRICA

SPONSORED BY THE MACARTHUR FOUNDATION FOR PEACE AND JUSTICE

Managing Editor: Dr Eureta Janse van Rensburg
Editors: Professors Johan Hattingh, Heila Lotz-Sisitka and Rob O'Donoghue

This book is printed on acid-free paper (Mondi Status 80gm^2 uncoated woodfree smooth bond)

Published by the Human Sciences Research Council Publishers
134 Pretorius Street, Pretoria, South Africa

First published 2002

Printed by Creda Press

ISBN 0-7969-2001-X paperback

Produced by comPress
http://www.compress.co.za

Distributed in South Africa exclusively by Blue Weaver Marketing and Distribution, P.O. Box 30370, Tokai, Cape Town, South Africa, 7966.
Tel/Fax: (021) 701-7302, email: blueweav@mweb.co.za.

CONTENTS

ACKNOWLEDGEMENTS

This monograph was produced by the Environmental Education Association of Southern Africa (EEASA) in partnership with the Human Sciences Research Council (HSRC), with funding from the MacArthur Foundation for Peace and Justice and WWF South Africa. The Managing Editor, Dr Eureta Janse van Rensburg, was assisted by three editors, Professors Johan Hattingh, Heila Lotz-Sisitka and Rob O'Donoghue.

LIST OF CONTRIBUTORS

Daniel Babikwa

Daniel Babikwa is a lecturer in the Department of Adult Education and Communication Studies at Makerere University, Uganda, and a Ph.D. student in Environmental Education at Rhodes University, South Africa. His academic interests include participatory methodologies in environment and development education and community-based development research. He comes from the Kalingala district of Uganda and holds a Masters in Development Studies from the University of Dar-es-Salaam in Tanzania.

Sara Freeman

Zimbabwean-born Sara Freeman has been a Waste Minimisation Officer in Durban Solid Waste, eThekwini Unicity, South Africa, for the past seven years. She obtained a Masters of Social Science (*cum laude*) in the School of Life and Environmental Sciences, University of Natal-Durban, and lectured part-time in Waste Minimisation, Recycling and Environmental Education for the Institute of Waste Management.

Johan Hattingh

Professor Hattingh is Associate Professor of Philosophy and Head of the Unit for Environmental Ethics at the University of Stellenbosch, South Africa, where he obtained his D.Phil. in Philosophy. His research interest is the role of ethics and values in environmental decision-making.

Heila Lotz-Sisitka

Professor Lotz-Sisitka is the Murray & Roberts Chair of Environmental Education at Rhodes University, South Africa. She holds a Doctorate in Education from Stellenbosch University for research into participatory educational materials development. She has been a leading figure in establishing environmental education in Curriculum 2005 and the National Qualifications Framework in South Africa.

Justin Lupele

Justin Lupele comes from Mansa in Zambia. A former high-school teacher, he is an Education Officer for WWF Zambia Education Project. He holds a Diploma in Agricultural Education and is researching participatory resource materials development towards a Masters of Education (Environmental Education) at Rhodes University, South Africa.

Lynette Masuku van Damme

Born in Swaziland, Lynette Masuku van Damme worked in the Swaziland NGO *Yonge Nawe* before taking up biodiversity education, cultural heritage and environmental education positions in conservation agencies in South Africa. Holder of a WWF Prince Bernard scholarship, she is conducting doctoral research into "technologies of transformation", indigenous knowledge and mutually beneficial partnerships between a Khomani San community and the Kgalagadi Transfrontier Park straddling South Africa and Botswana. She is the immediate past president of the Environmental Education Association of Southern Africa, and recently took up the position of Director, Heritage, in South Africa's Department of Arts, Culture, Science and Technology.

Ndyebo Mgingqizana
Ndyebo Mgingqizana is a Waste Minimisation Assistant in Durban Solid Waste, eThekwini Unicity. Durban-born, Ndyebo holds a Bachelor of Social Science from the University of Cape Town and an Advanced Certificate in Development Management from the University of Durban-Westville, South Africa.

Martin Mkandawire
Martin Mkandawire is a Doctoral Research Fellow in the Institute of General Ecology and Environmental Protection in Dresden University of Technology, Germany. Martin was born and is resident in Mzimba in Malawi. He worked as a *State of Environment* Reporting System Specialist with Malawi's Capacity Development in Environment Project. His research interests include the effects of radioactive tailings from mining on aquatic systems, and environment and development policy in African contexts.

Tsepo Mokuku
Dr Mokuku was born and resides in Lesotho, where he lectures in Science Education to student teachers at the National University of Lesotho. He conducts consultancy work in science and environmental education. His research interests are curriculum development and community-based natural resource management. Tsepo holds an M.Ed. in Science Education from Wits University and a Ph.D. in Environmental Education from Rhodes University, South Africa, and is a member of the EEASA Council.

Kevin Moore
Kevin Moore is a South African who started his career in the National Parks Board as an information officer. He is currently Regional Social Ecology Manager (Coastal Parks) in the South African National Parks. He holds a National Diploma in Nature Conservation, field-guiding qualifications and a BA degree from UNISA in Communications and Industrial Psychology. He is currently studying towards an M.Ed. in Environmental Education at Rhodes University, South Africa.

Edgar Neluvhalani
Edgar Neluvhalani is from the Northern Province of South Africa, where he was an environmental education lecturer at a college of education before taking up the post of National Technical Advisor to the National Environmental Education

Project in the Department of Education. He has an M.Ed. from Rand Afrikaans University and is currently the lead Ph.D. Indigenous Knowledge and Curriculum researcher in the Rhodes University Environmental Education Unit in Grahamstown, South Africa.

Rob O'Donoghue
Professor O'Donoghue holds a Ph.D. in Environmental Education from Rhodes University, South Africa, where he is the Director of the Gold Fields Environmental Education Service. Born in Zimbabwe, he worked as a primary school teacher and environmental educator in the Natal Parks Board and KwaZulu-Natal Wildlife, where he conducted much of his research into environmental education and indigenous knowledge as social processes. His work is reflected in numerous publications in the *Southern African Journal of Environmental Education* and a body of internationally-used educational resources.

Leigh Price
Zimbabwean Leigh Price is an independent environmental education consultant who has advised the sugar industry, and coordinates the Speciss College-Rhodes University Environmental Education Certificate Course for General Educators and for Industry, Business and Local Government in Zimbabwe. She has an M.Sc. in Tropical Resource Ecology from the University of Zimbabwe and is registered for a Ph.D. in Environmental Education at Rhodes University in South Africa. Leigh's research interest is discourse analysis of environmental education texts in industry.

Vladimir Russo
Vladimir Russo is from Luanda, Angola, where he was an environmental journalist and President of the NGO *Ecological Youth of Angola*. He is currently a resource material developer for the SADC Regional Environmental Education Centre, and the General Secretary of EEASA. He is exploring his research interests – media and environment, and learning support materials – in an M.Ed. (Environmental Education) programme at Rhodes University, South Africa.

Mike Ward
Mike Ward is employed by the Wildlife and Environment Society of South Africa to manage the SADC Regional Environmental Education Programme. He holds a B.A. in Philosophy and Geography from the University of South Africa and is currently studying for a Masters in Environmental Science at the University of Lund in Sweden, with a grant from the Swedish Institute. His research interests are issues of power and knowledge in situations of ambivalence with particular emphasis on the sustainability of funded programmes.

Environmental Education, Ethics and Action in Southern Africa

Many environmentalists regard the 2002 United Nations World Summit on Sustainable Development (WSSD) with ambivalence. The fact that thousands come together from all continents to focus on poverty, prosperity and the ecological basis of life is a magnificent reflection of global consciousness. The questionable contribution of previous UN conferences to actual socio-economic and political change is the not-so-shiny side of the coin.

Hosting the World Summit in Johannesburg presents the southern African region with a unique opportunity to reflect on and share its particular environmental challenges and responses. This monograph makes use of that opportunity, as it opens up for your consideration some of the issues Africa's environmental practitioners have been grappling with since the United Nations Conference on Environment and Development (UNCED) of 1992.

Environmental Education, Ethics and Action in Southern Africa is a window onto environmental challenges in diverse African contexts. These contexts include those of Malawian officials and community leaders, new to multi-level governance, taking up the challenge of environmental management in villages and districts; of Ugandan small-scale farmers in partnership with non-governmental organisations (NGOs) trying to produce sustainably for household and export markets; and of government-civil society partnerships in South Africa, where the political transformation of the education system introduced a focus on environment and human rights in the national school curriculum. Other contributions from South Africa, Angola, Lesotho, Zambia and Zimbabwe further discuss contexts of environmental practice: industry reportage, environmental management, research, philosophy, ethics, the media, conservation, and seeking out indigenous knowledge.

What is there to expect from this collection of papers, besides diversity and African perspectives? The framework for the monograph is deliberately open-ended. Contributors have simply been asked to review developments in their fields of practice, in relation to the ten years between UNCED and WSSD, and to highlight and comment on progress and challenges. We were hoping for perspectives that would sharpen our understanding of the issues as we respond – in our respective professions – to the environmental degradation and socio-economic injustices that are so poignantly present throughout Africa.

Authors have responded by providing empirical reviews (such as the review of the development of materials mobilising indigenous knowledge for environmental learning in schools, by O'Donoghue and Neluvhalani), conceptual reviews (such as

Hattingh's analysis of the concept of sustainable development) and case studies (such as Lupele's case study of global influences on his practice as an environmental educator in Zambia). The case studies were written from within the authors' own practices, as educators, government officials, researchers, development workers, journalists, conservationists and industry consultants. The authority of their contributions is in the personal experience from which they write. The benefit to the scholars, researchers, students and educators who read and use these case studies and reviews is not only that they document important trends and issues in a decade of environmental practice in Africa. They also provide considerable substance for analysis and critical deliberation, for these interlocking contexts of environmental practice should be at the core of our endeavours to understand and respond to the challenge of sustainable development, both practically and conceptually.

As editors we identified in these papers inter-related themes which seem particularly significant in relation to the discourses surrounding WSSD. These themes relate to what Thomas Popkewitz calls a social epistemology – an understanding which places knowledge (of, say, environmental issues and risks), knowledge production (research, education, communications) and acting on knowledge (environmental management and policy-making, for example) within webs of strategic social relations, woven through *power* and situated in *history* and *context*. In such webs, relations between and within individuals (the district environmental officer or the writer of company environmental reports) and institutions (governments, industry, donors, NGOs) shape and constrain, and are in turn shaped and constrained by discursive practices, that is, more or less evident rules for what can be said and done, what can be changed, and what remains as is. This point is perhaps most explicitly made by Price, who argues that authors of company reports are constrained in what they can and cannot portray in the annual environmental report, in ways which make the public "disclosure" of such reports – one of the key environmental responses in industry – particularly ambivalent. The point is also illustrated by Babikwa, who portrays extension work in a Ugandan NGO being shaped by successive discourses of social amelioration and welfare, food security, ecologically sustainable development in an emerging market economy, and, most recently, financial sustainability as the NGO becomes a business in a donor-driven economy. In another context (South Africa's national conservation agency), a similar trend with associated tensions is evident in organisational shifts from conservation *sans* people to social upliftment, to partnerships for socio-economic development, to a corporate enterprise that has to achieve a public mandate within an increasingly privatised business framework (described by Moore and Masuku van Damme).

The relational dynamics that make up the social epistemology of "sustainable development" draw the *boundaries* or horizons of the *change* towards which many, in one way or another, advocate, make policies, form partnerships, implement projects, donate funds, communicate and educate. In relation to desired changes towards social justice and ecological sustainability, every contribution in this monograph reflects tensions, contestations and contradictions. The contexts they portray, while complex and diverse, are all characterised by uncertainty and ambivalence. Environmental education, sustainable agriculture, environmental reporting, implementing environmental policy, working with donor funding, designing waste management plans – all are sites of struggle.

Babikwa illustrates how the clash of contradictory agendas within organisations (NGOs, donor agencies and in fact "communities") can frustrate good intentions. Mkandawire's contribution illustrates the same, and adds the insight that reform initiatives are often ambivalent: Malawi's central government leads that country's process of decentralising environmental governance to local authorities, but also fails to fully accept, endorse and support the process. Russo's case study of the conceptual and practical challenges facing an Angolan NGO engaged in environmental reporting argues that Agenda 21's call for the mass media to promote public participation in sustainable development may be made – and interpreted – from within questionable assumptions about targeted messages and social marketing for behaviour change, which can but fail to achieve desired ends. Hattingh reviews the contradictory interpretations of the concept of sustainable development across ideologically diverse socio-historical contexts. He argues that while some interpretations constitute an ideology critique aiming to foster change, others do not. Writing from Lesotho, Mokuku argues that development processes in African countries often contribute to what he terms the "de-development" that followed colonisation and industrialisation on the continent. His paper, like that of O'Donoghue and Neluvhalani, further illustrates ambiguities associated with the historical appropriation, marginalisation and current re-appropriation (through, for example, critical research) of indigenous knowledge. Even the eThekwini Unicity's success story of sustainable development in the recycling industry (by Freeman and Mgingqizana) reveals paradoxes as street-dwelling cardboard collectors are being brought into the institutional fold.

Ward argues that our efforts to deal with the uncertainties of environmental risks, through scientifically derived environmental management models and tools, has the paradoxical result of preventing us from reducing the manufacture of risk. Lotz-Sisitka, reviewing the challenges of advocating environmental learning in schools within the ambiguously transforming institutional framework of South Africa's education system, suggests that we need to re-think our very understanding of how social change comes about – a point implicit in several of the case reviews of policy implementation. Drawing on social theorists Bauman and Beck, she notes that in an ambivalent world, society has ways of dealing with contradictions and tensions which simply absorb critique, rendering it "toothless", unable to shift life politics. Price's discourse analysis of ESKOM's *Environmental Report 2000* reveals contradictory discourses on sustainability simply "cancelling each other out" within a set of discursive practices (disclosure from a reputable energy company and contributor to development on the sub-continent) which leaves no room for challenge and little doubt that the global power company of 2001 is doing the best it can under the circumstances.

And it probably is, if one recognises ESKOM as a player in a historical, contextual and strategic relational network which determines its possibilities of agency. As the notion of social epistemology suggests, it is these relations and not individual agents or ideologies which shape the production, dissemination and reception of knowledge and thus the possibilities of such social practice as sustainable development.

It is for this reason that we draw attention to the way in which the contributions to this monograph bring to the fore some of the often unacknowledged rules, power plays and other social processes through which knowledge and meaning are produced, circulated and received in society. These largely unwritten rules, the often silent social processes and taken-for-granted institutionalised mechanisms of truth production, are constantly at work in the very language and models that we use to conceptualise environmental problems and what we regard as suitable responses. The contributions collected here provide a vantage point on such processes, from which one can probe, empirically, conceptually and ethically, the consequent social distribution of benefit and of environmental degradation and risk. From such a vantage point we may start to re-interpret and re-appropriate knowledge and knowledge production processes, including education, action and critique towards socio-ecological change.

As narratives of social affairs which foreground people and the practice of policy-making and implementation on the very uneven terrain of struggle for social change, the contributions to this monograph seek to deepen our insight into the multi-directional interplay between donors and receivers, policies and projects, experts and clients, colonisers and colonised, and between individual members of rural communities who, as Babikwa puts it, "may look similar in their poverty", but have diverse needs and aspirations as unique nodes in complex historical, socio-cultural webs.

CHAPTER ONE

On the Imperative of Sustainable Development: A Philosophical and Ethical Appraisal

Johan Hattingh, South Africa

Introduction[1]

One of the most remarkable things about the concept of sustainable development is that it has become part of a small set of words like democracy, justice, fairness, equity and transparency that almost everyone seems to support, and almost no one seems to reject. As such, the concept of sustainable development has become part of a common vocabulary among those who are concerned about the impact of human activity on the ecological basis of our existence. In spite of the misgivings of some commentators[2], sustainable development has become strongly associated with a moral imperative that apparently no one can ignore or reject without having to provide very good reasons for dissent. And yet, while the *term* sustainable development has become widespread in recent times, there is little indication that a clear global consensus has also emerged about the *content*, the interpretation and the implementation of this moral imperative.

In this paper, I discuss four different interpretations of the moral imperative to promote sustainable development that have emerged since the early 1970s. Although the differences between these interpretations have only been clearly recognised in philosophical and ethical terms over the last decade[3], it is not possible to provide an overview of the emerging battlelines in debates about sustainable development by restricting oneself to developments since the 1992 Earth Summit. In fact, current debates about the meaning and implementation of sustainable development are still informed by various philosophical and ethical interpretations formed before 1992.

In the discussion of each one of these four interpretations, I shall briefly survey the historical and philosophical context within which that interpretation has emerged. Each will also be appraised in terms of its implications, effects, strengths and weaknesses. Attention will be drawn to the fact that none of these interpretations is ideologically neutral. Working with the critical conception of ideology as the utilisation of meaning in the service of asymmetrical power relations (Thompson, 1990), two of the sections will be concluded with an indication of how a particular interpretation of sustainable development can have the effect of establishing, justifying or maintaining relationships of domination and exploitation. Two further sections, however, will conclude with references to the significant role that the concept of sustainable development can play in ideology critique.

This is done with a view to show that the highly contested character of the concept of sustainable development can be ascribed to its ambivalent ideological dimensions: it can either function as an ideology, or as a critique of ideology (which of course in itself can be interpreted as an ideology, in so far as it serves the interests of those engaged in ideology critique). Furthermore, the objective here is to show that an agenda of radical ethical questioning and ideology critique can be used to find a position for decision-making and actions within debates about the meaning and implementation of sustainable development.

Sustainable development as a green agenda of nature conservation

A first glimpse of the ideological character of the different interpretations of the meaning and implementation of sustainable development is evident when we consider them within their historical and philosophical contexts. Achterberg (1994a; Eckersley, 1992) points out that one of these historical contexts is that of a realisation in the 1970s amongst Western nations that industrialisation and exponential growth in resource consumption and human population seriously jeopardise the continued existence of a safe, healthy, clean and diverse environment. Characterised as it is in terms of a crisis or a turning point for humanity, this realisation is articulated in various reports, principally The Club of Rome's *The Limits to Growth* (Meadows *et al.*, 1972) and *A Blueprint for Survival* (Goldsmith *et al.*, 1972).[4] In order to overcome this crisis and to ensure the survival of society, various structural adjustments to the economy and to social life were proposed in order to attain a state of equilibrium in which material growth is halted, although expansion in services enabling a higher quality of life in terms of education, cultural activities, experience of nature or enjoyment of leisure time was deemed to be in order.[5] With his proposals for a steady-state economy, Herman Daly (1973)[6] became one of the most articulate proponents of a zero-growth economy.

In the wake of Garret Hardin's essay *The Tragedy of the Commons* (1968)[7], theorists like Heilbroner (1974) and Ophuls (1973 and 1977) proposed highly controversial solutions to the environmental crisis, which continue to find sporadic support. These entail a virtual if not *de facto* suspension of democracy in which problem-solving is left to a centralised government and its bureaucratic structures. From this perspective, solutions are sought through structured planning based on expert knowledge, science and technology. It can be granted that proposals such as these are motivated by the sobering realism that people will not willingly give up a way of life, particularly when it entails the enjoyment of relative privileges. However, they very soon encountered the criticism that theirs was an authoritarian, top-down approach to the implementation of sustainable development (Eckersley, 1992). Participation of people affected by these solutions was not taken seriously, and alternative forms of knowledge, such as indigenous knowledge systems, were ignored (Jacobs, 1999).

An example of an alternative approach to the implementation of sustainable development from that time can be found in *A Blueprint for Survival* (Goldsmith *et al.*, 1972:30) in which four main pre-requisites for a sustainable society (conceptualised as a state of equilibrium) were articulated:

- a minimal disruption of ecological processes;
- maximum conservation of resources and energy;
- a population in which only the losses are replaced;
- a social system in which the individual does not feel limited by the first three conditions, but instead enjoys them.

Such a social system was envisaged to consist of decentralised, self-sufficient communities in which people worked close to home, governed themselves and developed their own personal and communal identities, formed values and ideals in which they could take pride, and could be happier than they would have been in the anonymous mass existence of a centralised metropolitan life.

In so far as this implies a transformative, if not revolutionary, agenda of an alternative lifestyle in which current patterns of production and consumption are challenged, and the ideal of minimal impact on supporting ecosystems is taken seriously, this interpretation has much in common with most of the other notions of sustainable development discussed below. However, it has been criticised for its utopian character since it serves an agenda of maximum environmental protection and nature conservation – the flipside of a total suspension of all material growth in the economy. It has also been criticised for having highly untenable implications for those living in developing countries.[8] A global policy of a zero-growth, steady-state economy would confine those living in developing countries to the trap of a highly skewed and unjust distribution of the world's resources, with no hope of ever changing the material basis, or substantively improving the quality, of their lives. As formulated by Feenberg (1979, cited in Eckersley, 1992:16), seeking a solution to the environmental crisis in a state of equilibrium "leads to a politics of despair that would freeze the current relations of force in the world – and with them the injustices they sustain – as a condition for solving the issue of survival".

An alternative, more realistic interpretation of sustainable development can be given in terms of ecological fit, i.e. staying within the carrying-capacity of supporting ecosystems (Achterberg, 1994a). Here, ecological sustainability is neither incompatible with the use of natural resources on a level higher than one of minimum impact, nor is it incompatible with a certain measure of material growth in the economy. In short, then, sustainability does not equal minimum impact on nature.

With this in mind, it becomes possible to unmask the ideological character of arguments for sustainable development that appeal to a state of equilibrium with minimum impact on ecosystems and maximum conservation of nature. Such arguments can justify maintaining and perpetuating a neo-Malthusian lifeboat ethic[9] in which the advantages of the affluent are protected (in the name of nature conservation), while a death sentence is pronounced on the poor (Eckersley, 1992) by not allowing them to utilise nature directly to secure their livelihoods.

Arguments for environmental protection to reach a steady-state economy are largely something of the past, but arguments for minimum impact on natural resources and maximum environmental protection are still prevalent today. Their potential to impact negatively on the opportunities of the poor to use natural resources directly to survive is real. As Shiva has confirmed repeatedly (e.g. 1988,1993, 2002), the direct use of nature by the world's poor may be their only chance of survival, and this need not entail use patterns that exceed the carrying-capacity of supporting ecosystems. On the contrary, there are numerous examples of the direct use of nature by poor people that are based on indigenous knowledge systems and display an ecological fit.

Problems like these indicate that the satisfaction of human needs, in particular those of the poor, as well as the notion of material growth within the economy, can and should form part of the core meaning of sustainable development. This is exactly what was emphasised in one of the other historical contexts within which the notion of sustainable development emerged during the 1970s.

Sustainable development as a social and economic agenda of needs satisfaction

The second historical context within which the notion of sustainable development has emerged is a series of United Nations conferences about environment and development, the first of which was held in Stockholm in 1972, and which also includes the Earth Summit held in Rio de Janeiro in 1992. A number of reports are closely associated with these conferences and have substantively informed them. The most important of these are the *World Conservation Strategy* (IUCN *et al.*, 1980) and *Our Common Future*, also known as the *Brundtland Report* (WCED, 1987).

In this context, the poorer countries of the world rebelled against Western preoccupations with the natural environment and the environmental policies to protect it. Where the notion of sustainability within the first context predominantly called for *limits to physical growth*, the call from within this second context was rather for *development (in particular of the poor) within the physical limits* of the supporting ecological systems. It was emphasised that our responses to environmental concerns should never be at the cost of the legitimate aspirations of the poorer nations of the world to overcome poverty and to reach a standard of living that is comparable to that of the richer nations. Accordingly, development in the sense of satisfaction of human needs, as well as in that of intra-generational justice, i.e. equitable access to the natural resources of the world, were seen as preconditions for sustainable development (Achterberg, 1994a).

It is within this second context that the *Brundtland Report* (WCED, 1987:43) formulated the well-known definition of sustainable development that has received global support: This definition states that "sustainable development is development that meets the needs of present generations without compromising the ability of future generations to meet their needs". What is often overlooked is the emphasis the *Brundtland Report* placed on the following:

- the concept of needs, in particular the essential needs of the world's poor, to which overriding priority should be given;
- the idea of limitations imposed by the state of technology and social organisation on the environment's ability to meet present and future needs.

This interpretation of sustainable development, however, seems to under-emphasise the imperatives of nature conservation and environmental protection, and fails to provide clear guidelines for environmental protection when the interests of humans and nature are directly in conflict with one another. This stems from the substantive anthropocentrism[10] of *Our Common Future*, and the so-called weak[11] notion of sustainable development that follows from it.

In terms of anthropocentrism, nature is considered to be of value only in so far as it can be utilised as a resource for humans. The intrinsic value of nature, i.e. the value that it has independent of its use value to humans, is ignored.[12] Similarly, environmental problems are predominantly conceptualised as management problems: they can be overcome through better management and/or better technology. This is clearly borne out by the fact that the only concept of limitation explicitly acknowledged in *Our Common Future* is that of the state of technology and social organisation. As such, this concept of limitation also reinforces what has become known in environmental economics as a weak conception of sustainable development.

According to this weak notion, sustainable development has been achieved if the *overall stock of capital* has been maintained over time. This means that natural capital (resources) can be exchanged with human and financial capital without compromising the ideal of sustainable development itself (Norton, 1992). As such, the weak notion is based on the assumption of the inter-substitutability of different forms of capital. The sacrifice of a natural resource could thus be justified indefinitely by (and balanced or traded off against) what is gained in terms of financial capital (expressed in monetary terms), or in terms of human capital (gains in, for example, infrastructure such as roads and buildings, or advances in science and technology). Weak sustainable development, therefore, only provides protection to parts or aspects of nature if it pays to do so, i.e. if its continued existence (conservation) could be traded for a higher price in monetary or human terms than it would otherwise fetch in an open market of commodities or services.

This kind of cost-benefit analysis characterises the dominant thinking about environmental policy and decision-making globally, and stands in stark contrast to the strong notion of sustainable development. Strong sustainable development entails the maintenance of *natural capital* over time. This implies that there are certain circumstances under which gains in financial and human capital cannot compensate for the loss of nature.

From this perspective, it becomes evident that a weak notion of sustainable development, legitimised as it is by its anthropocentric emphasis on meeting human needs (in particular those of the poor), can serve the ideological function of justifying and perpetuating the exploitation of natural resources far beyond the limits of the carrying-capacity of supporting ecosystems. Such unsustainable exploitation, the argument goes, is in order as long as it is balanced by gains in financial and human capital. As such, a weak notion of sustainable development cannot be used effectively to challenge patterns of production and consumption that exceed the limits of supporting ecosystems, but have a high financial or human value, for instance by producing a large number of jobs.

However, increases in financial and human capital cannot always balance or compensate for the loss of nature.[13] A stronger, more robust approach to sustainable development is needed in order to acknowledge the notion of the intrinsic value of nature, and at the same time overcome the dichotomy between the satisfaction of human needs and environmental protection.

Sustainable development as an integrated agenda of caring for the community of life on earth

An articulation of this stronger, integrative approach to sustainable development can be found in *Caring for the Earth: A Strategy for Sustainable Living* (IUCN *et al.*, 1991), the follow-up to the *World Conservation Strategy*. In this report, the ideals of sustainable development are refined into a strategy for *sustainable living*. Mention is also made of a strategy for a *sustainable economy*. The publication focuses on quality of life, and not only on survival issues. Holding onto the notion that an activity is sustainable if it can be maintained indefinitely, sustainable development is defined as improvement in the quality of human life, so far as it is possible within the boundaries of the carrying-capacity of the ecosystems on which it is dependent (IUCN *et al.*, 1991:7). A sustainable economy would be the result of sustainable development. Such an economy would keep its natural resource base intact, but could continue to develop by adapting to change and by improvements in knowledge, organisation, technical efficiency and "wisdom" (see Achterberg, 1994a:29).

Elaborating on the vision articulated for sustainable development in the *World Conservation Strategy*, nine principles are proposed in *Caring for the Earth* that should form the ethical platform of sustainable living. These are:

- respect and care for the community of life (an ethical principle that defines a duty of care for other people and for all forms of life, now and in the future);
- improving the quality of human life;
- conserving the vitality and diversity of the Earth;
- minimising the exhaustion of non-renewable resources;
- keeping within the carrying-capacity of the Earth;
- changing personal attitudes and practices, in accordance with an ethic for sustainable living;
- enabling communities to care for their own environments;

- forming a national framework for the integration of development and conservation;
- forming a world alliance to implement sustainability on a global scale.

Caring for the Earth (IUCN *et al.*, 1991:12) notes that these principles, values and duties are not new and that they have been articulated in many of the world's cultures and religions for centuries. They also reflect many of the statements that have been made at United Nations conferences and in reports on the need for equity, the participation of all stakeholders in decisions impacting on their well-being, conservation of nature, and economic efficiency as prerequisites for sustainable development. In short, these principles reflect an essential support for the principle of respect for life in general, emphasising the importance of nature and ecosystems that also include human life. The challenge for sustainable living from this perspective is therefore not to justify it, but rather to help individuals and nations apply it to concrete actions and practices. Accordingly, the bulk of *Caring for the Earth* is devoted to the suggestion and explanation of a large number of actions that should be taken within different contexts "if we are truly to care for the Earth" (IUCN *et al.*, 1991:65).

From the perspective of *Caring for the Earth's* integration of the concern about needs satisfaction and respect for the community of life, the Earth Summit of 1992 and its attendant documents (the *Rio Declaration* and *Agenda 21*) represent a step backwards[14] to a strong anthropocentric interpretation of sustainable development. In fact, in its emphasis on people in the concern for sustainable development, as well as for the eradication of poverty and injustice in the distribution of the world's resources, the Rio Summit and its declarations do not add anything new to what was articulated in the *Brundtland Report* of 1987.

This does not mean that the Earth Summit of 1992 was a failure. It did register a remarkable moral consensus among the political leaders of the world about the importance of sustainable development, as well as a commitment, if only in principle, to a demanding programme of action and policy formulation for the 21st century. According to Achterberg (1994a), the Earth Summit also confirmed a commitment to a political ethic of egalitarianism in terms of which all people should be accorded equal care and respect, regardless of the nation, culture or generation in which they find themselves.

The philosophical position of *Caring for the Earth* could thus be interpreted as an effort to establish a convergence between the perspective of sustainable development (nature conservation and environmental protection) that has emerged from industrialised countries, and the emphasis on development as needs satisfaction that has emerged from the poorer nations of the world. In its emphasis on respect for the entire community of life, the strong anthropocentric position of the *Brundtland Report* is challenged. At the same time, in its strong emphasis on improving the quality of human life, in particular by addressing the problems of skewed and unjust access to the world's natural resources, the strong nature-centred position of conservationists in industrialised countries is also criticised.

But what would it entail to implement these ideals in real-life contexts? On this question, *Caring for the Earth* only provides general, abstract answers, although one of its principles (namely changing attitudes and behaviour) may be pointing us towards exploring the notion of sustainable development as a radical political and ethical agenda of intellectual and institutional transformation.

Sustainable development as a radical political and ethical agenda of transformation

Sustainable development has always been closely associated with an agenda of social, institutional and intellectual transformation. In its first articulations in industrialised countries, it challenged dominant patterns of production and consumption, calling not only for a change in lifestyle and social organisation, but also for a fundamental reorganisation of the format of societal governance and the mode of material production. Similarly, from the perspective of the poorer nations of the world, sustainable development not only called for a more equitable distribution of the wealth of the world amongst everyone living at present, but also expanded the notion of justice to include future generations. Also, in *Caring for the Earth*, a revolutionary paradigm shift in our ethical perspective is proposed, in which concerns about the well-being of humans are embedded within respect for the community of life, without negating the moral imperative of addressing the needs of the poor and the destitute.

Taking a closer look at this transformative agenda, we can discern the underlying idea that incisive structural adjustments on different levels will be required in order to achieve sustainable development. Within industrialised societies, it was realised that mere adjustments in individual lifestyles and values would not in themselves establish a state of ecologically sustainable development. What was required, rather, was a fundamental reduction of consumption (implying an "economy of enough") and a drastic collective change in lifestyle, linked to structural changes in social institutions and government (Achterberg, 1994b:143). At the same time, rich as well as poor nations realised that issues of *international justice*, that is, the distribution of wealth between those living now, and joint decision-making about the matter, should receive structural and institutional attention as essential prerequisites of sustainable development.

The idea of sustainable development, however, challenges us also to think and act transformationally with regard to the self and self-realisation. This point is neatly captured by Eckersley (1992:17–21) when she observes that the environmental crisis at its deepest level is not one of physical limits to growth and distributional issues alone, but also a crisis of culture and character. What she means by this is that the environmental crisis of our time challenges us to the point of questioning the very notion of material progress, and lamenting the social and psychological costs associated with the dominance of instrumental rationality. Included among these costs are alienation, loss of meaning, the coexistence of extreme wealth and extreme poverty, welfare dependence, dislocation of tribal cultures, and the growth of an international urban monoculture with a concomitant reduction in cultural diversity (Eckersley, 1992:17-18).

Turning this depressing cultural diagnosis into a positive point, Eckersley (1992:20) proceeds to argue that this situation has been perceived by a new breed of ecopolitical theorists active in the 1980s and 1990s[15] as an opportunity to emancipate us from the total of the inherited ideas, beliefs, values, and social practices that underlie the environmental crisis. In this regard, the focus falls in particular on challenges to dominant thinking in three inter-related areas, namely human needs, technology and self-image. This leads to critical inquiries into the structure of human needs and the appropriateness of many modern technologies. Summarised in the words of Eckersley (1992:20):

> It is no longer considered adequate merely to challenge, say, the site of a nuclear power plant, freeway or chemical industry, or merely to insist on better safety devices or pollution filters. Instead ... [we have] to draw attention to the more fundamental question: to what extent do we really need these kinds of energy sources, these means of transport, these industries and technologies ...?

The transformational agenda of sustainable development leads to similar far-reaching questions about who we are and what we live for on this earth. What is at stake here is the whole conception of, and relation between, humanity and the world – the material, cultural and spiritual worlds – which leads us further to the central and eternal question: what is human life? (Castoriadis, 1981, cited in Eckersley, 1992:20).

That questions like these have been formulated amongst emancipatory theorists and political activists from the 1960s onwards does not reduce their contemporary currency. Still relevant today, they are being discussed in increasingly widening circles of people concerned with sustainable development. It is a pity, however, that most of these questions are not taken seriously, let alone deliberated, by the dominant centre of society. A clear case in point is the decision by the American government not to ratify the Kyoto Protocol of 1997 to reduce global warming.

These questions are too often and too easily dismissed in certain circles as the ranting of a lunatic fringe that has to be marginalised as effectively and as quickly as possible (witness responses to the demonstrations at the Seattle meeting of the World Trade Organisation in 2000). It is, therefore, significant that many of the questions and issues that constitute the radically transformative agenda of sustainable development have been taken up since the beginning of the 1990s by a revitalised civil society across the globe that functions alongside and independently of the state. Thus the imperative to promote sustainable development merges with the emancipatory and critical agendas of an astonishingly wide array of social movements and organisations concerned with the issues of feminism, peace, third-world aid and development, HIV/AIDS and the like. Furthermore, these organs of civil society, constrained as they are by prejudicial media coverage, lack of funds, fragmentation etc., move to a large extent one step ahead of theory, and can therefore inform theorists in finding ways of "overcoming the destructive logic of capital accumulation, the acquisitive values of consumer society, and more generally, all systems of domination (including class domination, patriarchy, imperialism, racism, totalitarianism, and the domination of nature)" (Eckersley, 1992:21).

Conclusion

What follows from this discussion of different interpretations of the moral imperative to promote sustainable development? In the first place, it is evident that none of these interpretations is neutral; rather, they represent ideological positions in so far as they justify and promote the interests of certain sectors of global society. The fact that these interests clash and are often mutually exclusive is a clear indication that any interpretation of sustainable development functions as a set of normative ideas. Such a set of normative ideas can function as guidelines for personal actions, and as a baseline in terms of which governments, industry, commerce, consumers and citizens can be held accountable for their actions.

The serious problem, however, is that there are different interpretations of the moral agenda of sustainable development, and this can lead to different models of accountability. In this regard, it can be useful to look at this diversity from the perspective of ideology critique, and to identify a number of critical questions that should be asked of each interpretation. Formulated in general terms, some of the central questions that need to be asked are: whose interests are served by adopting this or that agenda of sustainable development? Whose power is served and through which mechanisms? And who or what stands to win or lose in which ways from adopting this particular agenda of sustainable development, rather than that one? Are new forms of dependency created by adopting this or that interpretation of the agenda of sustainable development? Are new forms of domination and exploitation created, or are we in the process of creating conditions that slowly but surely push back domination and exploitation in the world? With questions like these, we can start to come to grips with the ethical implications of the different concepts of sustainable development discussed above.

What has also emerged from the discussion above is that the different interpretations of sustainable development provide ideologically loaded answers[16] to fundamental value questions (see Achterberg, 1994a:36) such as:

- What is so important that we should strive to maintain it forever? Is it expansion in material growth; consumption; survival; needs satisfaction; quality of life; the flourishing of life on earth; the diversity and abundance of life on earth; or the ecological basis of life in general?
- With a view to whom or what should we pursue the sustainability of this valuable something? Do we do it for the sake of nature, or the sake of people; do

we do it for the sake of the rich or that of the poor; or for the sake of the whole of the community of life?

- How should we pursue sustainability? From a centralised position in a top-down manner with experts and science and technology; from a participative position in a bottom-up manner in which consensus, as well as indigenous and cultural knowledge systems, plays a large role; or with a combination of these approaches as circumstances and context dictate?
- How would we know that we have moved nearer to or further away from sustainable development? Do we make use of financial indicators alone; do we use wider and more comprehensive economic indicators to assess costs and benefits; do we use indicators from social and political life – for example the satisfaction that people experience from work, or what they can identify with as individuals and groups; do we use indicators from nature – for example the behaviour of indicator species to detect almost imperceptible changes in the climate of a region; or do we combine all of the above? And exactly how do we determine the threshold values that should apply to any set of indicators that is chosen, and whether they are exceeded or not?

In the last instance, it can be maintained that the imperative to promote sustainable development is a moral appeal to all of us to think clearly and anew about the answers that we personally give to the four fundamental value questions above.

From this perspective, the agenda of sustainable development is one of radical and critical questioning of ourselves and our motives, of the social bases of our actions, and of the implications and effects of these actions on others: people, future generations and other members of the community of life. If this point is missed, however, sustainable development could become just another entry in the current list of ideologies in the service of the *status quo* that leave the world, with all its risks and injustices, much as it is.

Endnotes

1 This is a substantially expanded version of Hattingh (2002) and Hattingh (2000).

2 Achterhuis (1994) refers to the lie of sustainability. Jickling (1999) is concerned about the moralistic and deterministic overtones of much of the propaganda for sustainable development. Patterson (1998) sees the concept of sustainable use as a green mask used by industry and governments to justify and continue the ruthless exploitation of natural resources.

3 See Eckersley (1992), Achterberg (1994a, 1994b), Jacobs (1999).

4 First published as vol. 2, no. 1 of *The Ecologist*, 1972.

5 Ideas like these can be traced back to John Stuart Mill's *Principles of Political Economy* (1848:752–756). See Achterberg's (1994a:22–23) related discussion.

6 Other proponents of this "survivalist school" are Hardin (1968 and 1992), Ophuls (1973 and 1977) and Heilbroner (1974); see Eckersley (1992).

7 He argued that freedom in the unregulated commons would bring ruin to all, and therefore that only "mutual coercion, mutually agreed upon by the majority of the people affected" would put us in a position to overcome the crisis of survival.

8 These implications form a sub-class of the generic criticism that exponents of a steady-state economy are misanthropic, i.e. that they deplore humankind.

9 Such a lifeboat ethic, heavily criticised as it has been for its Malthusian assumptions and implications, has been explicitly formulated by Hardin (1968 and 1992).

10 However, it is true that certain phrases, particularly in Chapter 2 of this report, suggest a marginal acknowledgement of the intrinsic value of nature.

11 The notion of "weak sustainability" is explained later. Its "weakness" refers to the limited protection that it offers to nature that has no use value to humans.

12 This is predominantly true of stronger versions of anthropocentrism. In versions of enlightened anthropocentrism, the value of the independent existence of nature can be acknowledged, although this is often done on the basis of the instrumental value of nature for purposes of recreation, or for the satisfaction gained from knowing that wild nature still exists somewhere on earth.

13 Achterhuis (1994:202), for example, argues that the recommendations of the *Brundtland Report* could, "after a short-lived profit, lead to long-term accelerated destruction of both humanity and nature".

14 The stronger, more robust interpretation of sustainable development of *Caring for the Earth* is not so much motivated by ecocentric sentiments as it is by a form of weak (or enlightened) anthropocentrism – a position that is widely accepted amongst moderate environmental ethicists. See Norton's (1984) positive evaluation of enlightened anthropocentrism.

15 These theorists include Murray Bookchin, John Rodman, Theodore Roszak, Christopher Stone, Arne Naess, Bill Deval, George Sessions, Kirkpatrick Sale, John Dryzek, Karen J. Warren, and Val Plumwood.

16 Jacobs (1999) conceptualises the different answers that can be given to these value questions in terms of a number of "fault lines" that can be discerned in the concept of sustainable development.

CHAPTER TWO

Integrating Economic Development, Social Justice and Ecological Sustainability:
A Case of Sustainable Development in the Waste Industry, eThekwini Unicity, Durban

Sara Freeman and Ndyebo Mgingqizana, South Africa

Introduction

The South African government's vision for environmental management, as stated in the *White Paper on Environmental Management Policy in South Africa* (Department of Environmental Affairs and Tourism [DEAT], 1999), is a society that is in harmony with its environment through an integrated and holistic environmental management system that aims to achieve sustainable development. To achieve sustainable development, an environmental management system needs to address:

- the quality of people's lives, and their daily living and working environments;
- fair access to land and natural resources;
- the integration of economic development, social justice and environmental sustainability;
- more efficient and sustainable use of resources;
- public participation in environmental governance;
- the custodianship of our environment (DEAT, 1999:13–14).

Sustainable development is however a difficult concept to pin down because it means different things to different people, depending on where they sit on the political, economic or social spectrum (see Hattingh, this volume). In South Africa a close relationship between economic activity, the environment and development is assumed, and this needs to be taken into account in the promotion of sustainable development. This chapter describes a case from eThekwini Unicity (Durban), where a waste management initiative is recognising this relationship and simultaneously addressing job-creation and income-generation, capacity-building towards greater social equity, and ecological sustainability through recycling and the reduction of waste.

Sustainable development in the South African context

Perhaps the most frequently quoted of many contested definitions of sustainable development, is that from *Our Common Future* (WCED, 1987:43): "sustainable development is development that meets the needs of the present without compromising the ability of future generations to meet their needs". This definition is extended to imply promoting development that does not consume resources faster than they can be replaced by natural processes or technology. It also combines concern for the environment with social and economic issues, to ensure that human health and natural resources are not compromised (DEAT, 1999).

O'Riordan (1998:99) observes that while in "the South, for the most part, survival and the maintenance of social order seem to be ... prime motives, and sustainable development [is] ... regarded as essentially a Northern agenda [receiving] short shrift from politicians and community leaders", in South Africa the significance of sustainability is recognised "as an index to environmental and social health, linked to economic opportunity". While South Africa is a developing country that needs growth and development to improve the quality of people's lives through the creation of new jobs and wealth, it also needs to use environmental resources in a sustainable way, and to consider *all* these factors as relevant in the realisation of sustainable development.

O'Riordan describes South Africa as being:

> entwined in a triple helix ... a combination of economic redistribution, social justice via democratisation and shared opportunity, and environmental protection linked in part to the public health. The primary drivers are economic renaissance for the poor and improving the rights and liberties of those formerly oppressed by apartheid. The economic development effort is geared to the positive redistribution of income, to job training, and to employment creation (O'Riordan, 1998:101).

The concept of sustainable development has been adopted locally at municipal level through the acceptance of the Local Agenda 21 (LA21) mandate. Agenda 21 was conceived as a global action plan for socially, economically and environmentally sustainable development extending into the 21st century, at the 1992 United Nations Conference on Environment and Development. In 1994, the then Durban City Council accepted the LA21 mandate and ratified the implementation of the mandate as a corporate responsibility. Durban's LA21 process is seen as a tool for producing an environmental management strategy focusing on sustainable development and community participation in planning, decision-making and capacity-building to ensure a healthier, cleaner, safer, and greener city (Durban City Council, 1998).

In 1995, in response to the LA21 mandate and in anticipation of central government setting targets and promulgating laws to move away from traditional "end of pipe" waste disposal solutions and towards integrated waste management systems, Durban Solid Waste (DSW) set up a Waste Minimisation Office. DSW is a waste management and disposal contractor for the eThekwini (Durban) municipal region. The objective of the new office was to divert waste from landfill sites through the promotion of waste minimisation, recycling and related job creation in the greater Durban area. To achieve this objective, the office secured funding for and established a number of sustainable formal and informal buy-back and drop-off recycling centres in conjunction with the private sector, formerly unemployed street collectors and community groups.

Based on this experience, which we share in this chapter, we would suggest that while national policy-making in South Africa regularly "genuflects to the constraints of sustainable development" in the process of transition to a developed and democratic society, and "sadly development often continues apace at the expense of the environment" (Blowers, 1997:846–47), this programme can satisfy the broader South African vision of sustainable development with its interlinked social, economic and ecological concerns.

However, a caution must be noted that the "relationship between global economic pressures, sustainability, and the maintenance of a democratic civil society remains a close call in South Africa" in a "drama that frames the sustainability transition in a myriad of paradoxes and tensions" (O'Riordan, 1998:99), as our experience in this initiative bears witness.

Recycling in South Africa and the search for sustainability

Like sustainable development, recycling has to be viewed in economic, ecological and social contexts.

In terms of economics, recycling in South Africa at present is not generally subsidised by either the industry or government grants and green taxes, as is done elsewhere, particularly in Europe and North America. In addition, the vast majority of households are poor and unable or unwilling to pay rates to the local municipality for services provided. For example, one particular suburb in the eThekwini municipal area contains some 65 000 households, of which only 900 pay rates (DSW, 2000). In households in first-world countries, ratepayers pay for being able to put out items to be collected for recycling, and are often also faced with punitive "pay as you throw" schemes whereby they pay by weight for all waste sent for disposal instead of recycling.

Recycling in South Africa is therefore subject to market forces and vastly fluctuating prices paid on either local or international markets for recyclable items. It is thus important for the profits of any recycling scheme to outweigh the costs, considering both operating and environmental costs. In our experience, essential factors to consider when gauging the economic, social and environmental viability of a proposed recycling scheme and to ensure its sustainability are:

- markets for the sale of the waste items, which must be located fairly close to the source of the recyclables and sorters separating them out of the waste stream, in order to cut down on transport costs, and which must pay sufficient monies to cover operational and overhead costs like labour, handling and rental for premises;
- suitability of the recovery scheme for the particular type of waste stream being sorted in the search for recyclable items;

- location of recycling centres, which will determine whether it is to be a drop-off point (in an affluent area) or buy-back centre (in areas with many unemployed people, to support income generation with the exchange of recyclables recovered for cash);
- quantities and quality of recyclable items available in the waste stream, which will determine how much is paid for each batch of recovered items relative to the effort that must be expended to find, sort and bale those items;
- technology appropriate to the conditions of the area, where specialised skilled labour and financial resources may not be available to operate and maintain a sophisticated recycling scheme;
- sufficient secondary markets, which will absorb the items made from the recyclable waste, for resale to the public (DSW, 2002).

In line with these considerations, DSW decided to support existing and to establish new buy-back and drop-off centres which could serve as pilot schemes for the further establishment of centres in eThekwini Unicity municipal area. From early on we recognised the need to foster public-private partnerships (between the local authority, the recycling industry, informal collectors and residents) in order to establish and support these centres. There was also a need for fundraising to supply infrastructure, to create a significant number of temporary and permanent jobs for the unemployed, and for training at all levels to ensure upward mobility in line with the city's LA21 programme.

Public-private partnerships of this kind are supported by the South African government, as reflected in the *White Paper for Local Government,* Department of Constitutional Development (DCD, 1998), and the *White Paper on Environmental Management Policy* and *Agenda 21* (DEAT, 1999). Specifically it is stated (DEAT, 1999:115–16) that the effectiveness of governance structures is influenced by the capacity of civil society to work together with government. In keeping with international trends and *Agenda 21*, the government will affirm, strengthen and/or establish partnerships with interested and affected parties including:

- non-governmental organisations and associations;
- affected communities and community-based organisations;
- organised labour;
- business and industry, and/or industrial associations, with special attention to the needs of small, medium and micro-enterprises;
- other government departments in all spheres of government.

Most importantly, the policy adds the need to establish structures, mechanisms and processes to foster public participation to provide good governance, advise and inform government, and achieve co-ownership for development (DEAT, 1999). We believe, as the other partners in the eThekwini initiative did, that capacity-building through training is an important mechanism to foster such participation (see also Freeman, 2000).

The project: establishing buy-back and drop-off centres

Two key pilot centres established by DSW were used as models for the creation of further recycling centres in the eThekwini Unicity area. Both were established as public-private partnerships.

Brook Street

The Brook Street buy-back centre was formally launched in August 2000 in Warwick Junction, a major commuter hub in central Durban. It was set up to establish an organised system of cardboard collection for informal collectors in the Warwick and Central Business District (CBD) areas, in an attempt to end cardboard stacking at street corners, and to provide less exploitative economic opportunities than those which existed at the time for informal collectors in the area. The project was started by the DSW Waste Minimisation Office which facilitated the process of getting all relevant parties together to negotiate the setting up of a centre which would serve the interests of all partners.

Before negotiations took place, DSW conducted a waste stream survey in both the CBD and Warwick Junction to determine the potential location of the centre and to establish the recyclable with the highest volume and value currently being traded to the recycling agents on street corners. This exercise helped determine the viability of the project, as well as the probable number of informal collectors in and around the proposed centre. A number of informal collectors were found by interviewing recycling agents.

Following the results of the survey, the relevant parties came together to negotiate the logistical issues of the implementation of the buy-back centre idea. The following groups were invited to participate in the negotiations:

- Babs Waste Paper (BWP, a paper recycling agent which had been dealing with informal collectors in the area for 15 years);
- Mondi Recycling (which supplied equipment for the project and purchased cardboard and paper from BWP for recycling);
- the Informal Trade Department of the City Council (which regulates informal trade within the eThekwini Municipality);
- the Warwick Junction Urban Renewal Project (which provided and fenced the land for the buy-back centre and spearheads urban renewal in Warwick Junction);
- the Self-Employed Women's Union (SEWU, a union for informal collectors in the Warwick Junction area).

An agreement was struck in 2000, in terms of which BWP would operate the Brook Street buy-back centre on a lease whose terms included the nominal rent of R10 per month from the council, while continuing to buy paper and cardboard from informal collectors and selling it in turn to Mondi Recycling. It was agreed that SEWU would provide a representative to run the centre on a full-time basis under the supervision of BWP, and be paid a weekly salary by the agent during the pilot period. It was also stipulated that this person would enter into a profit-share agreement with BWP once s/he had gained enough experience in the business. BWP and Mondi Recycling provided the necessary start-up equipment for the project, such as a cash float, a shipping container to store paper and cardboard, a portable scale, an office container for administration, two trolleys for haulage of paper and cardboard, and signage boards inside the site. DSW, together with BWP, provided basic training to the SEWU representative and co-monitored the operation of the centre. The Informal Trade Department provided the legal agreement for the centre between the Council and BWP, and Warwick Junction Urban Renewal Project supplied the land and slab for the centre, free of charge. This partnership agreement also made provision for a six-month trial period during which the operation of the site was to be monitored.

During the pilot period the centre proved to be a success, collecting a monthly average of 30 tons of paper and cardboard which generated about R10 000 per month. Every month, R7 000 was paid to the informal collectors and the remaining R3 000 was split between BWP and the SEWU representative. After the pilot period other recycling companies, Collect-A-Can and Glass Recycling Services, entered into the partnership, providing storage containers for cans and glass respectively. These other recyclables contributed comparatively less volume (less than a ton per month) because of their low value among informal collectors.

Towards the end of 2001, a profit-sharing system between the SEWU representative and BWP was introduced, with BWP still providing the operating cash float. This new arrangement was partly necessitated by the fact that there had been a drop in the monthly volumes of recyclables at the centre to ten tons a month. Immediately after this change, average monthly volumes picked up again to the 15 to 20 ton range.

An extended survey was conducted by DSW in March 2001 among informal collectors throughout the CBD including Warwick Junction, Beachfront and Victoria Embankment, to assess the need for another buy-back centre and the effectiveness of the Brook Street centre. This survey indicated that Brook Street was under-utilised because of a number of factors. Firstly, 60 per cent of the informal collectors in the city were between the ages of 41 and 70 and not eager to walk distances longer than a kilometre to a centre carrying a heavy load. They preferred to sell to the trucks from the recycling companies on nearby street corners. The second reason cited for poor use of the centre was that most collectors did not have trolleys which could be used to carry their cardboard to the buy-back centre. Attempts are being made to provide multi-compartment skips as an alternative to the buy-back centre, in which informal collectors can store recyclables. The establishment of another buy-back centre in closer proximity to certain groups of collectors is also being considered.

There are two other issues relating to the area in which the Brook Street centre is located. First, the tendency of informal collectors to live on the pavements just outside of the buy-back centre, which negatively affects the image of the site. Second, traffic congestion at the centre's entrance, stemming from the fact that the surrounding area is a busy commercial district with limited parking, making it difficult for recycling companies' trucks to enter and exit the site.

Bellair

The Bellair drop-off recycling centre, part of the Bellair Garden Refuse Site on Bellair Road, was one of the first formal drop-off sites for recyclables established by DSW. It was formally opened in October 2000, after the previous drop-off facility had been closed for over a year because of problems experienced with informal waste reclaimers breaching security and continually invading the site to recover items of value from the waste stream.

The Waste Minimisation Office motivated to the planners of the new garden-refuse site to construct a separate, formalised, roofed drop-off area for recyclables on the site, together with a turning circle to control traffic entering and exiting the facility. Although several of the municipal garden-refuse sites already had containers in them for the collection of recyclables, this was the first centre where a separate area for recyclables was planned and constructed. The land allocated to the facility was council-owned, with DSW paying the rates and the monies for construction sourced from the capital budget. It was decided to confine the function of the centre to that of a drop-off facility, to avoid the problems experienced with informal reclaimers at the previous site.

For the first year after construction, the recycling centre was overseen by the Waste Minimisation Office, with various recyclers supplying containers for the collection of paper, cardboard, cans, glass, motor oil, plastics and scrap metals. As a service to both the public and the municipality, the recyclers removed the recyclable items from the site free of charge, selling them to cover their transport and handling costs. During that first year, records were kept of the amounts of recyclables being dropped off and sold. These figures were used to motivate for a concessionaire to be appointed through the council tender process to operate the recycling facility as a small business enterprise. The concessionaire would pay a reasonable rental to DSW, amounting to the approximate value of the paper and cardboard collected on site at that time.

RainGem Consulting won the tender to operate the recycling centre for 24 months from December 2001. The main selling point of its offer was the proposal to train a previously disadvantaged, unemployed person to run the centre for the first year, and then to enter into a profit-sharing arrangement with that person for the second half of his/her contract. The concessionaire was previously employed in the waste management and disposal industry, and had a sound knowledge of local recycling industry dynamics. In terms of managing the drop-off facility for recyclables, RainGem provided containers for the depositing of the recyclable items on site and arranged for the transport of these in bulk to markets for sale, while also publicising the centre and undertaking to educate the public on site.

The Bellair facility is extremely popular with the public, and it is estimated that some 900 vehicle owners visit the site monthly in order to drop off recyclables and garden refuse. Some residents travel a distance of up to 20km in order to use the site. In fact, the system has proved so successful that the concessionaire applied for, and won, a contract to undertake a number of scoping exercises to provide similar services at other garden refuse sites in the eThekwini municipal area.

Poverty Relief Fund pilot recycling centres

With the experience gained from Brook Street, Bellair and other centres[1], DSW decided to seek funding from donors to set up further pilot recycling centres in order to create further jobs and divert more waste from landfill sites in the region.

The process began in July 2000, with the drafting of a business plan seeking funding for infrastructure and equipment. Construction was based on the plans and costing of Bellair while the equipping and operationalising of the centres were based on both Brook Street and interviews with representatives of the local and national recycling industry. It was envisaged that the centres be built on council-owned land, with a motivation put forward to council for nominal rates and rental to be paid by the emerging entrepreneurs.

In December 2000, representatives of the national Department of Environmental Affairs and Tourism (DEAT) expressed interest in supporting the business plan through the Poverty Relief Fund. In January 2001, R3.7 million was allocated to the Association of Clean Communities Trust, the funding arm of DSW's education and waste minimisation division, to build and equip two pilot buy-back recycling centres and equip two scrap-metal buy-in shops.

The next step in the process was to find suitable, well-placed council-owned land for the sites, to draw up plans and to identify previously disadvantaged persons who could operate the facilities as small businesses, employ staff and create networks of informal collectors to supply the sites with recyclables.

Finding land proved to be the biggest obstacle to the project. Approval from the eThekwini Council for the site chosen in Westmead and for nominal rental and rates was only gained in December 2001, and, in the case of the North Coast Road Centre, as late as June 2002. From experience, in order for the recycling centres to be viable without the benefit of subsidies or green taxes, they had to be situated at the interface of a thriving (not declining) industrial/commercial area, and in close proximity to a residential area, the households and waste streams of which could eventually also be tapped for marketable recyclables. The entrepreneurs also had to be in a position to collect recyclable items directly from the majority of these outlets without having to pay for them, so as to generate sufficient income to cover their operating costs.

Two scrap-metal shops are currently in operation and construction has just been completed on the Westmead Recycling Centre, which will be handed over to the entrepreneur in mid-2002. It is envisaged that the construction of the North Coast Road Centre will be completed by August 2002.

The selection of entrepreneurs and construction contractors, as well as the purchase of equipment, has been facilitated through a tender process in line with eThekwini municipality's procurement policy, which favours the appointment of previously disadvantaged individuals and suppliers. Advertisements appeared in the local media calling for expressions of interest from individuals keen to operate and staff the centres. A total of 16 groups were invited to attend a workshop at which they were briefed in order to prepare and submit proposals and business plans. From this group, eight were short-listed for intensive interviews, based on their proposals and business plans. From the interviews, four entrepreneurs were selected to run the sites, with briefing and training meetings taking place every two weeks in order to assist them in drawing up workable cash-flow plans, and in the selection and purchase of equipment for each facility.

Obstacles to this process included the fact that these previously disadvantaged people had to obtain bridging finance in order to operate the businesses until profits could be realised. The funding supplied from DEAT was expressly reserved for capital and infrastructure costs. In the case of the scrap-metal shops, the entrepreneurs were tasked with financing the rental and setting up of services for their premises. This was a challenging prospect as rentals in suitable areas were generally high. For the recycling centres, the concessionaires were tasked with factoring rental costs into their cash-flow plans. All four had to provide proof of insurance for the equipment being supplied by the funding via the Trust. As first-time business owners and operators, all of them had difficulty securing funding from financial institutions. Assistance was eventually gained from small business lending associations and shareholders.

Training and education for sustainability

There is wide agreement among development practitioners that education has an important role to play in leading to development that is sustainable. This entails empowering individuals and groups to participate in sustainable development (see Freeman, 2000). *Agenda 21* devotes a whole chapter to education for sustainability in which it is stated that:

> Education is critical ... for promoting sustainable development and improving the capacity of the people to address environment and development issues ... for achieving environmental and ethical awareness, values and attitudes, skills and behaviour consistent with sustainable development and for effective public participation in decision making (Fien, 1996:2.27–2.28).

This is supported by the *White Paper on Environmental Management Policy in South Africa* (DEAT, 1999) which has identified seven interdependent strategic goals for sustainable development, one of which is: "[to] promote the ... education and empowerment of South Africa's people. To increase their awareness of, and concern for, environmental issues, and to assist them to develop the knowledge, skills, values, and commitment that we need to achieve sustainable development" (DEAT, 1999:46).

In terms of this concept of empowerment, from the inception of the Poverty Relief Fund project, the vision was that the informal collectors would eventually be trained as sorters/balers at each centre. The sorters would in turn be trained to become supervisors of satellite centres, and the supervisors would be trained to become entrepreneurs in their own rights. The training provided would not only concentrate on the provision of technical knowledge associated with the recycling industry, but also with business, financial and life skills. A problem identified with the previous buy-back and drop-off centres was the difficulty in capacitating informal collectors to become centre entrepreneurs or small-business owners.

It was envisaged that this training would be provided by the Waste Minimisation Office with assistance from programmes already in existence in the recycling industry, but it was agreed that additional funding be sought to finance this training. Monies from an international donor were secured with the agreement that individuals from other buy-back and drop-off centres would be invited to attend the training sessions for the purposes of capacity-building and sustainable job creation in the waste industry.

Conclusion

This chapter proposes that by addressing environmental, social and economic matters, the establishment and support of recycling buy-back and drop-off centres such as those in the eThekwini Unicity municipal area serve as mechanisms for achieving sustainable development in South Africa.

This initiative recognises the desire of informal collectors to make a living from the recyclables they recover from the waste stream and trade for cash, but also provides them with a capacity-building mechanism to eventually own and operate their own buy-back or drop-off centres as small businesses for profit. In supporting the establishment and operation of these centres, the local authority benefits from reducing waste at landfill sites in line with government policy, and from building the capacity of and providing employment to members of previously disadvantaged communities, thus reducing the burden on the social security and municipal infrastructure systems. Valuable resources are also saved in the process. The recycling industry also gains in that it can readily access large and concentrated quantities of recyclables from the waste stream for recycling at a relatively low cost.

What is also key to the initiative on a national level is that these experiences now form part of a User Friendly Guide on Recycling being developed by the DEAT to inform local authorities in the process of establishing and supporting local waste minimisation and recycling schemes as part of their legal mandate.

The ultimate goal of this process has been to attempt to reduce the power differentials between the recycling industry and the informal collectors in order to promote social, economic and environmental equity. It is proposed that this will lead to sustainable development which, in the words of Durban's LA21 policy, will assist in promoting peace, prosperity, safety and security in the eThekwini Unicity (Durban City Council, 1998).

Endnote

1 For example, the Merebank community and municipal drop-off sites, and the Umlazi, KwaMashu, Cato Manor, Tongaat, Quarry Heights, Inanda, Lindelani and Ntuzuma buy-back centres.

CHAPTER THREE

Environmental Management: Expertise, Uncertainty, Responsibility

Mike Ward, South Africa

Introduction

In the follow-up to the World Summit on Sustainable Development (WSSD), many experts will be called upon, or will position themselves, to implement and interpret the suggestions emerging from the Summit and its associated processes. In doing so, these experts will consciously or unconsciously fulfil different roles and adopt a range of strategies in carrying out their tasks. They will thus continue the trend of specialist guidance being provided by consultants and advisors in environmental management, particularly since the 1992 Earth Summit.

This chapter seeks to make more explicit some of the "knowledge contexts" (Bauman, 1987) or "theories of knowledge" (Beck, 1999) that may underpin the specialist decisions taken and guidance given, and their possible implications. More specifically, the knowledge contexts of modernity and post-modernity and theories that emphasise the linearity and non-linearity of knowledge will be considered. Linked to this is a discussion on the roles of experts as "legislators and interpreters", drawing on Bauman's (1987) book of the same title. The chapter includes an example of how these considerations may enable a more informed use of environmental management tools; in this case, tools associated with systems analysis and computer modelling. The main argument is that many experts (whether self-proclaimed or acknowledged) are operating in situations of ambivalence and uncertainty. In such situations, I will argue that issues of power and knowledge, as they relate to ethics and responsibility, need critical consideration.

Theories of knowledge – linear and non-linear

In his book *World Risk Society*, Beck (1999:124) makes a distinction between linear and non-linear theories of knowledge. A linear conception of knowledge emphasises the possibility of, and striving for, one coherent knowledge system. Underpinning this conception are strong assumptions of universality, foundation, homogeneity, monotony and clarity. Unknowns, or conflicting knowings, are played down as "not-yet-resolved-but-in-principle-resolvable" imperfections of knowledge (Bauman, 1993:8). Unawareness, defined by Beck (1999:127) as both inability to know and unwillingness to know, is thus marginalised within linear theories of knowledge, as a temporary nuisance or failing. The flipside of the marginalisation of unawareness is the idealisation of positive knowledge and certainty. Closely linked to this certainty is an absence of doubt about the universal grounding of knowledge. Thus, the cumulative creation of new knowledge based on the consensus of universally recognised practitioners operating from within universally sanctioned institutions becomes the defining property of a linear theory of knowledge. This understanding of knowledge has much in common with Bauman's portrayal of modernity as being about "conflict-*resolution*, and about admitting no contradictions except conflicts amenable to, and awaiting resolution" (1993:8).

It is the disbelief that everything is ultimately knowable and that conflicting knowings are no more than misunderstandings awaiting resolution that supports and is supported by theories of post-modernity. Those moments when the visions and hopes of modernity seem misguided and unattainable provide a glimpse of what has been described as the post-modern condition. This condition stems from the recognition that "there are powers behind the *plural* forms of life and plural versions of truth which would not be *made* inferior, and hence would not surrender to the argument of their inferiority..." (Bauman, 1987:141, emphasis in original).

Non-linear theories of knowledge accept unknowns as well as plurality, dissent and conflicting knowledge claims as central and inevitable components to understanding knowledge construction, deconstruction and reconstruction processes (such as scientific research and environmental management planning). The implications of having to make decisions and knowledge claims in the face of known unknowns (those areas where we are aware that we lack knowledge) and unknown, unintended consequences (those areas in which we have yet to recognise the implications of our actions), become significant in a non-linear theory of knowledge. There is a recognition that unknown, unintended consequences and suppressed unknowns do not erase endangerment but in fact intensify it. In addition, unintended consequences and unknowns do not stop or disappear as a result of unawareness (unknown and suppressed) but are favoured by it.

The focus in a non-linear theory of knowledge thus shifts to the "unclear, uncooperative and oppositely polarised networks of people and coalitions acting on knowledge" (Beck, 1999:125). In such situations, grand narratives, or universal knowledge claims, dissolve into processes of decision-making and power relations. A plurality of interest groups, none powerful enough to claim an objective superiority for the knowledge it represents, engage in processes of coalition formation around contested certainties and unknowns.

Berger and Luckmann, articulating the link between knowledge and power, write:

> The confrontation of alternative symbolic universes implies a problem of power – which of the conflicting definitions of reality will be "made to stick" in the society ... will depend more on the power than on the theoretical ingenuity of the respective legitimators (Beck, 1999:99).

Subsequent writings on knowledge/power relations have emphasised the plurality of positions and the play of dependencies between these positions. Foucault, for example, set out to "render apparent the polymorphous interweaving of correlations" (1975:58) by focusing on the multiplicity of component discourses which resulted in subtle transformations rather than unequivocal processes of social change (Hajer, 1997:47). In doing so, he criticised "the historical practice which sought to understand history in terms of causality" (Hajer, 1997:47). Similarly, Bauman describes situations in which the different actors and institutions operate within "spaces of chaos and chronic *indeterminacy*, a territory subjected to rival and contradictory meaning-bestowing claims and hence perpetually *ambivalent*" (Beilharz, 2001:178). This ambivalence "confounds calculation of events and confuses the relevance of memorised action patterns" (Beilharz, 2001:282). Bauman argues that in such situations "heuristics of pragmatically useful "next moves" displaces the search for ... certain knowledge of deterministic chains" (Beilharz, 2001:179).

This indeterminacy and ambivalence implies "that we cannot be sure what is going to happen, and ... neither know how to behave, nor can [we] predict what the outcome of our action will be" (Bauman, 2001:57). This in turn "generates acute discomfort and creates in the sufferers a craving for the security which only the practical control, or intellectual awareness, of probabilities may bring" (Bauman, 1987:10). Thus the lack of certainty and control generates the demand for (and is capitalised on by) a special group of people, or so-called experts, who (claim to, or are perceived to) possess the knowledge that enables them to discern order beneath superficial randomness, so that the indeterminate and ambivalent can be made predictable, clear and manageable.

Expert strategies

To the extent that these experts are unable to reduce this underlying ambivalence, they may adopt a number of roles or strategies. One strategy is characterised by Bauman (1987:4) as the "legislator" role which "consists of making authoritative statements which arbitrate in controversies of opinions and which select those opinions which, having been selected, become correct and binding". This strategy depends for its success on being carried out in situations of uncontested clarity (relatively rare) or on the expert being able to command enough power to silence or exclude any dissenting opinions (again rare in situations of institutionalised plurality). A second strategy outlined by Bauman is that of an "interpreter". This role consists of translating statements made by one interest group so that they can be understood within the system of knowledge of another interest group. "Instead of being orientated towards selecting the best social order, this strategy is aimed at facilitating communication between autonomous interest groups" (Bauman, 1987:5). There is often a strong assumption within this strategy, based on Habermas's notion of communicative action (Habermas, 1970), that different interest groups will aim at reaching agreement or consensus on a shared understanding of the situation as a basis for coordinating their activities. If, however, one emphasises Habermas's notion of strategic action (action based on self-interest), the possibility for consensus is substantially reduced. This is further complicated by the problem of where one draws the boundaries between communities of actors. What may be termed communicative action between a coalition of interest groups is very often a strategic action *vis-à-vis* other coalitions of interest groups.

Given these situations of ambivalence and indeterminacy, interests and power, those people cast as experts often create or turn to methods or tools that are themselves supposed to offer possibilities of clarity and certainty. One such set of tools that is gaining popularity in interdisciplinary environmental courses as well as in the broader environmental management arena is associated with systems thinking or systems analysis (see Anderson & Johnson, 1997; Haraldsson, 2002). These tools include causal-loop diagrams, behaviour-over-time graphs and computer modelling in programmes such as Stella[1].

It is claimed that these tools offer a precise set of rules that reduce the ambiguities and miscommunications that can develop when we talk with others about complex issues. There is a risk, though. In the following section, I will argue that the pre-established conventions that underpin these tools (like many others that we use to generate and express our understandings) influence our conceptual boundaries and create areas of silence that have ethical implications.

To illustrate how systems thinking and associated models may lead us into dangerous places, I draw on a task (2001) I undertook with fellow students of environmental management at Lund University. Our assignment was to produce an extensive Municipal Solid Waste Plan for the city of Schnitzellburg, that was "environmentally and economically feasible for the city in terms of recycling the material and/or incinerating the waste". In the process, we would draw on systems thinking and associated models.

Systems thinking and models

Much of the introductory literature on systems thinking focuses on the interconnectedness of all things. "The major point that systems thinking makes is that everything and everyone is interconnected in an infinitely complex network of systems" (Anderson & Johnson, 1997:vii, 20). A system is defined as "a group of interacting, interrelated, or interdependent components that form a complex and unified whole" (Anderson & Johnson, 1997:vii, 2). Based on these insights, systems thinking stresses the circular rather than the linear, focusing on "closed interdependencies". It has a precise set of rules that, it is claimed, reduce the ambiguities and miscommunications that can develop when we talk with others about complex issues. It offers visual tools that open a window onto our mental models, translating our individual perceptions into explicit pictures that can reveal subtle yet meaningful differences in view point (Anderson & Johnson, 1997:vii, 21). Given all of these attributes, systems thinking and the associated tools of causal-loop diagrams, behaviour-over-time graphs and computer modelling programmes are put forward as appropriate tools for an environmental science that aims to be interdisciplinary in character and holistic in approach (Haraldsson, 2002:4).

Given the complexity of many of the environmental issues that people in different contexts are facing, there is certainly something comforting about a system of thinking that promises to clarify the causal interconnections between things and to reduce ambiguity. In so far as systems thinking enables and supports communication between different disciplines, representation of interconnections and cycles, and the understanding of events and linear processes as being parts of larger systems, it has enormous value. However, we should be wary of uncritically adopting the underlying assumptions of modernity and linear theories of knowledge. There is a tendency to seek clarity through simplification, suppression of unawareness (unknown and suppressed) and a strong assumption of unambiguous causality.

"A system is a group of interacting, interrelated, or interdependent components that form a complex and unified whole" (Anderson & Johnson, 1997:2). Given this definition and the current focus on holistic thinking and interconnectedness in environmentalism, the system boundaries seem very broad indeed. This recognition of the scope of and relational complexity within systems serves as a useful corrective for the positivist tendency to narrow and simplify through the identification and isolation of variables. On the other hand, however, it is not possible to consider everything and so we do tend to draw boundaries around our areas of study either consciously or unconsciously.

Thus, in the assignment on the Schnitzellburg Waste Plan, we could look at the following causal-loop diagrams:

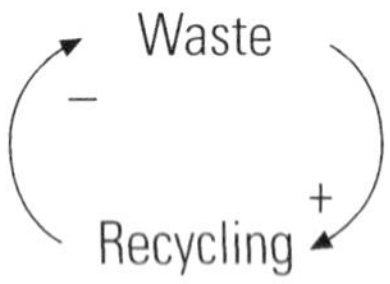

or

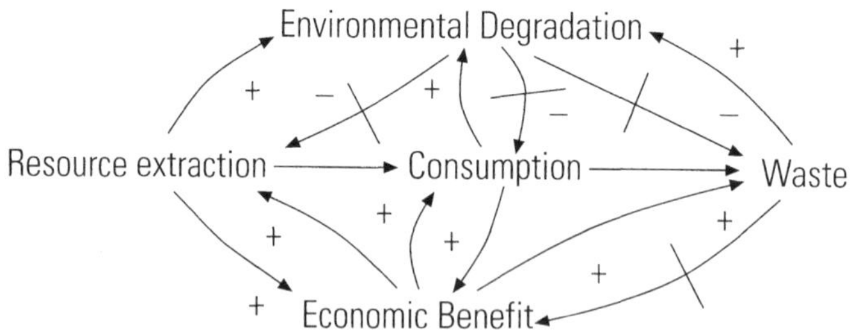

The first loop considers simply the role of recycling in reducing the amount of waste produced, while the second includes issues of resource extraction, consumption and waste in the context of economic benefits and environmental degradation. We are still missing any representation of unequal access to economic benefits and exposure to environmental risk. If, as a consultant, I decide to focus on the first loop, I cannot justify my decision on the grounds that there is some system boundary around waste and recycling. And if I limit myself to the second causal-loop diagram and exclude issues of equity, then I must take full responsibility for having done this. These boundaries are not predetermined. They are political decisions of inclusion and exclusion. In this case, they condemn the societies and ecosystems that are being exploited for the benefit of the Schnitzellburg residents to continued exploitation. What I must do is decide whether considering consumerism or inequality is important in a particular context, with all the implications of this decision. This inclusion or exclusion of links and impacts is not something inherent to systems thinking or its tools. In fact, systems thinking focuses on making these links more apparent. But practitioners can undermine this strength by claiming that it enables us to draw boundaries for which we have no responsibility.

There are other dangers of uncritically adopting systems thinking models. In the process of using models and tools to make apparent to ourselves and others how a particular object of study is to be viewed or understood, we engage in processes of re-presentation that ignore some things, put some other things in new contexts, and identify what is significant and what is not. In other words, in collaboration with our models and tools, we systematically deviate from, and in this sense distort, the original object of study as we perform analytical work on that object and re-present it to an audience. In doing so, we inevitably assert a kind of interpretive privilege over the original object of study and thereby deny it the right to speak for itself. The object's meaning has been appropriated by the model (and hence the model developer) used to represent it. While this may seem unimportant when talking about atoms or trees, it is critical when talking about people and societies. As causal-loop diagrams and Stella models are being applied to people, they become the objects of study. Complex emotional beings are reduced to quantifiable variables and in the process we risk making decisions about how best they can be manipulated and controlled. The appropriation of the meaning that other people and societies construct for themselves, and the objectification of these people and societies, have been recurrent themes in the human sciences (Mulkay, 1991) and need to be discussed by environmental management students and experts.

Focusing on the tools used in systems thinking, we run the risk of ignoring the complexity, historical contingency and fragility of the practices that we invent to make sense of ourselves in the world, and of the world around us. In doing so, we run the risk of developing a tool envy that blinds us to the multiple ways we have created (and are creating) to express and generate our understandings. Since the 1800s philosophers and practitioners in a wide variety of disciplines[2] have been responding to positivist perspectives on knowledge by exploring social constructions, understanding (as opposed to explanation), complexity, context, ambivalence, risk, uncertainty and the need for critique. By locating systems thinking within a broader field of intellectual and practical struggle around knowledge contexts and theories of knowledge, we may be in a better position to recognise the weaknesses and enhance the strengths of the tools currently used to support systems thinking.

One of the claimed strengths of systems thinking and its associated tools is that it offers a way of reducing the ambivalence and uncertainty in situations.

As Bauman (2001:57) remarks, the words

> ambivalence, ambiguity and equivocality … convey the feeling of mystery and enigma; they also signal trouble, whose name is uncertainty, and a dismal state of mind, called indecision or hesitation. When we say that things or situations are ambivalent, what we mean is that we cannot be sure what is going to happen, and so neither know how to behave, nor can predict what the outcome of our action will be. Instinctively or by learned habit, we dislike and fear ambivalence, that enemy of security and self-assurance.

Linked to these words on ambivalence is Beck's (1999:6) portrayal of modern industrial society as a risk society:

> The concept of "world risk society" draws attention to the limited controllability of the dangers we have created for ourselves. The main question is how to take decisions under conditions of manufactured uncertainty, where not only is the knowledge-base incomplete, but more and better knowledge often means more uncertainty.

One would expect that given our dislike of ambivalence, risk and uncertainty, we would shy away from the manufacture of risk. But by enabling us to represent complex systems in which we are embedded, models create the appearance that risks may be determined. Using Stella, we can model and compute the statistical probabilities of disaster striking, as well as the probable size of the disaster. This has the strange effect that if the threat of risk or disaster is horrifying, its modelling and calculability is – in defiance of logic – consoling (Bauman, 1993:201). It seems that a degree of comfort is brought about by the illusion of control over destiny. Perhaps what is needed in conjunction with systems thinking and modelling is a good dose of humility to see that we really do not have the capacity to understand, let alone control, all of the risks we manufacture.

Conclusion

The assignment on the Schnitzellburg Waste Plan is an example of how systems thinking and Stella models may lead us into ever-more dangerous places. If asked to write an essay on a Waste Management Plan that is "environmentally and economically feasible for the city in terms of recycling the material and/or incinerating the waste", most students would raise questions about resource exploitation, social injustice, etc. However, given the complexity and difficulty of including these dimensions of the issue in a causal-loop diagram or Stella model, I suspect that most would simply not bother to include these variables. There is an urge to produce a model that will allow us to predict and control the waste system in Schnitzellburg. And thus simple, perhaps even elegant models are produced that ignore issues of uncertainty, and create the illusion that we are in control and can continue business as usual with a few superficial adjustments.

As Bauman (1995:80) warns:

> Ambivalence always generates the demand for expertise; lack of orientation attracts experts in road-finding. Experts tend to promise a once-for-all, secure escape route from uncertainty, but what they offer in practice is a decision on how to "settle" the present dilemma without in the least reducing its ambivalence. We need them [experts] above all as authority; as someone we can trust because everybody trusts them, so that when accepting their advice we may be less tormented with doubts or guilty feelings than when we act on our own responsibility.

And so it is that as the council of Schnitzellburg reaches a crisis of ambivalence, it turns to consultants and experts who in turn grasp at what appears to offer certainty, namely modelling and computers. There is a very real risk here that rather than facing the manufactured risks of waste production (consumption, resource exploitation and social injustice) and recognising that we are out of control, we smooth things over with illusions of simplicity and certainty.

Similarly, if future experts are to be better equipped to exercise ethical responsibility in the post-Summit world, the assumptions underpinning the tools of the environmental management trade will have to be opened up for critical discussion so as to enable a more sensitive and critical application of systems thinking in environmental contexts.

Endnotes

1 High Performance Systems Inc., 1998. Stella is a computer modelling programme that enables modelling of systems through stocks, flows and converters. It is based on the ability to quantify the different variables.

2 Nietzsche writing in the late 1800s opens up a radical scepticism that forecasts the writing of Wittgenstein (mid-1900s), Foucault (late 1900s) and Derrida (current). Dilthey (1800s), writing on hermeneutics, also opened up space for focusing on understanding that was developed by Gadamer and more recently by Habermas in his critique of Gadamer.

CHAPTER FOUR

Decentralising Environmental Management in Malawi:
The Challenge of Capacity-building

Martin Mkandawire, Malawi

Introduction

Since the late 1980s many African, Asian and Latin American governments have launched decentralisation programmes aimed at establishing new frameworks for self-governance at sub-national and local levels (Bromley & Cernea, 1989; Furtado, 2001; Peterson, 1997). Decentralisation focuses on redressing the centre-periphery power and resource imbalances that often exist in developing countries. By putting more authority and decision-making power in the hands of local constituencies, decentralisation is an attempt at making social programming more responsive to local needs (Beyene, 2000). Thus, it has become an important dimension of governance and public sector reform programmes throughout the developing world, and is often linked to discussions on the role of improved governance in reducing poverty and achieving sustainable development (Bossuyt & Gould, 2000; Cheema and Rondinelli, 1983).

Malawi's national decentralisation initiative, starting in 1990, came to full throttle following the enactment of the Local Government Act in 1998 (Malawi Government, 1998). Administrative and political authority (functions, power, responsibilities) is slowly being devolved to a local level. This is expected to promote popular participation in governance and development, enhancing community participation in planning, decision-making and the co-management of resources.

The functions of environmental and natural resource management, monitoring and planning have been devolved by the Environmental Management Act (Malawi Government, 1996) to the District Assembly, which is the lowest level of authority. In response, the central government is deconcentrating institutions to put professional and technical know-how within easy reach of each community. It is hoped that decentralising environmental management in Malawi will improve efficiency, equity, development and resource management through greater participation in public decision-making. It is argued that public sector accountability and effectiveness would consequently also increase (Fox & Aranda, 1996). Efficiency increases with tapping the creativity and resources of local communities. Equity is promoted with more equal distribution and greater retention of benefits from environmental resources (Crook & Manor, 1998). It is also hoped that coordination and flexibility will increase among sectors involved in environmental action planning and implementation.

The process of decentralising environmental management is, however, proving difficult to both conceptualise and actualise at national and local levels. One contributing factor is a history in which central government was strongly inclined to retain authority over social programmes. This means that the majority of national and local officials are unfamiliar with the responsibilities that accompany the decentralisation of social service delivery. In the case of environmental management, this is exacerbated by low levels of technical environmental knowledge.

This chapter aims to highlight these and other challenges which the environmental management decentralisation process faces in terms of capacity-development, environmental education and training.

Decentralising environmental governance

Highly centralised administration has been the norm since Malawi's independence in 1964. The country is divided into three administrative regions, which are subdivided into 24 administrative districts. The central government was represented by a District Commission, which coordinated district programmes. The Commission had no control over other sectors because each sector at the district level was responsible to its respective department or ministry at national level. A local government that sat at district council level, parallel to the Commission, had limited powers. Policy, planning, implementation, management and even financing of programmes were centralised.

The new norm outlined in the National Decentralisation Policy (Local Government, 1998) and National Decentralisation Act (1998b), is a merged District Commission and Council to form the District Assembly headed by a mayor or chairman as a political leader, and chief executive as a civil servant. The District Assembly structure was to combine all sectors at the district level into one administrative unit where resources and technical support would be managed. Power was meant to be devolved to local political and traditional leaders who would take responsibility for forming local by-laws and policies, assisted by sectoral committees drawn from governmental and non-governmental technical support agencies. The thrust has been popular representation in local government through elected political bodies at district, area and village level. Hence local government would mirror central government.

The success of the decentralisation of environmental management lies in a range of factors involving the transfer of power, authority and resources away from the central level, and the creation of legal and institutional frameworks (Schacter, 2000). It is important to secure the appropriate decentralised institutional anchorage for environmental management.

The National Environmental Action Plan (1994) gave rise to the national framework for environmental management under which the Environmental Affairs Department was established. The department was formally constituted through the National Environment Policy (1996) and Environmental Management Act (1996) which also guide environmental management, especially relating to Community-based Natural Resource Management (CBNRM). Elements of CBNRM are included in recent revisions of policies and acts in Wildlife, Fisheries, Forestry, Water, Irrigation, Land, Agriculture and Energy sectors.

The Environmental Management Act (Malawi Government, 1996) requires that District *State of the Environment* Reports be produced every two years and compiled into the National *State of the Environment* Report, a summary of which should be placed before parliament every year. The District *State of the Environment* Report provides guidance to select the priority environmental "hot spots" for district planning. At the same time, it becomes the impact-monitoring tool for the implementation of the District Environmental Action Plans. District Environmental Action Plans should be produced every five years and compiled into the National Environmental Action Plan. The latter becomes the basis for updating the National Environmental Support Programme and policy, and is placed before the National Assembly every five years. The Environmental Action Plans are a vehicle for budgetary requisition to parliament, and an instrument to source donor support. The National *State of the Environment* Report becomes a parliamentary monitoring tool of the impact of the National Environmental Action Plan. The long-term vision of decentralisation is that the District *State of the Environment* Reports and Action Plans will become the main tools for district and national level environmental reporting and planning.

The Local Government Act (1998) places the onus on the District Assembly to attend to a number of environmental responsibilities, and to draw up plans for social, economic and environmental development in the form of a District Development Plan. Article 22 of the Environmental Management Act requires the Assembly to take charge of all decentralised services and activities, and to assist national government to protect natural resources and prevent environmental degradation. The Act stipulates the responsibility of the District Assembly to prepare District Environmental Action Plans and District *State of the Environment* Reports, coordinate activities in the protection, conservation and sustainable use of the environment, undertake extension services and disseminate environmental information for public awareness. The Assembly may appoint a number of committees at local government, ward, area or village level (e.g. a Health and Environment Committee) and has the power to establish environmental by-laws as required.

The implications are that the District Assembly is the lead agency responsible for ensuring that the environment is mainstreamed into the process of District Development Planning, and that environmental considerations become an integral part of Area and Village Action Planning. As we will see below, this is where capacity-building is needed most.

Article 20 of the Environmental Management Act (Malawi Government, 1996) stipulates that each district appoints an Environmental District Officer who is a public officer and a member of the District Assembly. The Environmental District Officer assists the District Assembly with support from the Environmental Affairs

Department. He or she retains relationships with this national department, and reports to its Director. Policy formulation and national planning, environmental inspection and law enforcement, research, the establishment of standards, training and curriculum development, facilitating Environmental Impact Assessments (EIAs), and international representation, are the responsibilities of the Environmental Affairs Department. The Environmental District Officer thus reports to the Director of the Department on related matters including professional standards, EIAs and training, and is visited by the Department from time to time as part of a continuous monitoring and capacity-development process.

At the same time, the Environmental District Officer is also directly accountable to the Chief Executive of the District Assembly, and under the Assembly's direct control for day-to-day operations. This "dual-boss" system does give rise to implementation difficulties. The District Officers have the lion's share of responsibilities and are from time to time at a crossroads in carrying out their diverse delegations and reporting duties.

At national level, the National Committee on Environment advises the government on policy. The National Focal Point represents sectors that plan and implement environmental-related policies. At district level there is a corresponding District Focal Point for executing the plan and policy, namely the District Environment Sub-committee. District Environment Sub-committees are anchored under a District Executive Committee. At sub-district level, there is an Area Development Committee and a Village Development Committee composed of traditional leaders and frontline staff (extension workers). The sub-district levels are supported by Focal Points at district and national level.

The National Environmental Focal Point is expected to communicate to the sectors and provide backstopping assistance to the District Officers. It is to meet bi-monthly to review and report at national level on the developments in the various sectors of the districts. Its Steering Committee meets every three months to assess the developments. Unfortunately, a lack of harmony between policies and acts in the various sectors makes this committee's task difficult. Furthermore, environmental management is still viewed as an extra mandate which is not accepted for financing under the core programmes of a sector, but only as external projects, or as the responsibility of the Environmental Affairs Department.

The environmental management decentralisation process

The district development planning process calls for a situational analysis, resulting in a Socio-Economic Profile. The contribution from environmental management will be the *State of the Environment* assessment which will be reported as the District *State of the Environment* Report. This includes coverage of the state, pressures, and response indicators to monitor trend patterns, and aims to highlight issues and identify the geographic locations of environmental "hot spots", by sector. It becomes the environment section of the Socio-Economic Profile. As the Profiles lead to the formation of District Development Planning Frameworks, the District *State of the Environment* Report leads to the formation of the District Environmental Action Plans. The latter is therefore considered to be the environment chapter of the District Development Plan, and includes mitigating actions to deal with key environmental concerns.

The environmental elements of the District Development Plan (i.e. the District Environmental Action Plan) will be implemented through environmental micro-projects. These will be funded either under a proposed "green window" of the District Development Fund or through other funding. Under a programme approach, donors could choose within a particular geographic location to fund a "time-slice" of the Action Plan. The District Development Fund stipulates monitoring and evaluation requirements which will allow feedback to the planning cycle and provide for forward planning. The same applies to projects implemented under the Action Plan, and monitoring of micro-project impacts feeds into a District *State of the Environment* Report.

Hence, the *State of the Environment* Report and Environmental Action Plan should be an integral part of the district planning cycle, and an obligation of local government. In this way, Malawi is to achieve the long-term goal of incorporating environmental issues into its new District Development Planning process.

As part of its management strategy, every sector produces an annual report. The sectoral *State of the Environment* Reports are extracts from these routine reports and plans of activities. The chapters of the *State of the Environment* Reports include Human Environment, Agriculture/Land Husbandry, Forestry, Fisheries, Water, Wildlife and Climate. Compilation of the sector's environmental status at village level produces the Area *State of the Environment* Report and Environmental Action Plan. These are compiled into District *State of the Environment* Reports and Environmental Action Plans, which are in turn incorporated into the National *State of the Environment* Report, produced and read in parliament every two years, and into the National Environmental Action Plan, presented every five years.

The success of the *State of the Environment* Reports and Environmental Action Plans would seem to lie in the uptake of ownership of the process to various agencies, and their capacity to develop environmental indicators, monitor issues, report and advise. The process benefits from and uses existing sector statistics and Socio-Economic Profile data. The *State of the Environment* Report is a decision-making tool, intent on identifying environmental problems, and seeking out community-based solutions for action.

Links to national policy

There are close links between local and national government institutions, and policy documents that accompany the decentralisation of environmental management in Malawi. The District Development Plans are not only in conformity with national Development Policy and Plans (DEVPOL), but also feed into the latter. Based on the Socio-Economic Profile and development trends, the Plans define districts' visions, opportunities and options for development. This includes elements of infrastructure, social services, investment, land-use planning and production. In this process it should, as a matter of routine, also undertake environmental impact assessments of all proposed developments. Socio-Economic Profiles feed into the National Development Report on which DEVPOL is based. Similarly, District *State of the Environment* Reports feed into the National *State of the Environment* Report, which should be submitted to Parliament every second year. The Act thus empowers the district authority with a powerful tool with which to lobby for support and resources at the highest level.

Capacity-building for decentralised environmental management

It has been stated that the success of decentralisation depends on the capacity of local governments to make and implement policies appropriate to available resources, and in accordance with the needs of the region's citizens (Sundaram, 1999; United Nations Development Programme, 1993; World Bank, 1989). Decentralisation of environmental management requires new or enhanced capabilities to analyse the needs of local populations, set priorities, undertake strategic planning and manage resources. The transfer of authority must be accompanied by a long-term strategy that supports local officials in developing the necessary capabilities (Rohdewohld, 2000).

In Malawi, the need for capacitating local government is intensified by the fact that for a long period, in the previously centralised administration, local government was mainly equipped with skills to execute orders from above. Consequently, at local level, there is little capacity to plan more effectively based on the ability to analyse the situation and environment, and implement such plans in the face of new or unexpected challenges. Hence, the decentralisation of environmental management in Malawi strategically started with capacity-building of appropriate institutions.

Capacity-building is a highly elastic term that can be stretched to refer to a variety of processes, including training, formal and informal education, networking, technical assistance and investments in infrastructure. In the decentralisation of environmental management in Malawi, capacity-building is viewed as a process by which individuals, groups, organisations and institutions strengthen their ability to carry out environmental management functions and achieve the stipulated legal mandates. It is expected that once capacity is fully built in decentralised environmental management, there will be enhanced efficiency, effectiveness and sustainability in environmental and natural resource management.

In 1997, the Malawi government, with the assistance of the United Nations Development Programme (UNDP), launched the Capacity 21 Project. With the assistance of the Danish International Development Agency (DANIDA), the Capacity Development in Environment Project, the Karonga-Chitipa and Lake Malawi Lakeshore District Environmental Management Projects, and the Lake

Chilwa Wetlands and Management Project were also launched to facilitate decentralisation. These projects worked hand in hand with the Environmental Education, Training and Outreach Unit of the Environmental Affairs Department.

Training and outreach

Through the Capacity Development in Environment Project, the Malawi government developed a training manual to guide capacity-building for decentralisation of environmental management. The manual has five modules:

- the legal and institutional aspects of decentralised environmental management;
- the *State of the Environment* reporting process;
- the *Environmental Action Planning* process;
- the implementation of micro-projects;
- monitoring and evaluation.

The approach to training was informed by a vision that people should be empowered to participate in environmental management through problem-based learning, because adults learn more effectively when challenged with a pertinent problem to solve. Thus, the decentralised environmental management process was introduced through awareness-raising seminars and training workshops that used a Learning-centred Problem Discovery and Action-oriented approach. Participants were presented with a problem and required to come up with the underlying causes, possible solution and their own plan of action to solve the problem. The training process was extended over a year in which training workshops were followed by "work-away" sessions where participants were expected to apply their developing skills. The training programme had focused targets: directors and National Environmental Focal Points at central level, District Executive Committees and District Environmental Sub-committees at district level, area and village committees and community-based organisations at sub-district level.

The Capacity Development in Environment Project, in collaboration with the Environmental Education, Training and Outreach Unit of the Environmental Affairs Department, also launched an outreach programme, using a wide range of strategies of environmental briefs (consisting of fact sheets), newsletters, radio and television programmes, organised talks and focus group discussions. The outreach programme had a very broad range of targets including environmental managers and policy-makers in public and private positions, civil society members at all levels and traditional and political leaders.

Politicians and professionals

Members of Parliament were oriented to the process of decentralised environmental management through briefs and newsletters distributed to parliament, and through workshops. This was done to win political support. Members of Parliament were also expected to communicate the decentralised environmental management process to their constituencies. However, the members were challenged with massive needs in their constituencies and issues of environmental management were usually among their lowest priorities.

Among the professionals, directors of natural resource sectors were targeted for an orientation to the institutional and legal provisions and tasks associated with the decentralisation of environmental management. Following the training, the National Focal Point institution was formed to encourage interdepartmental interaction as well as interaction with the sectors' respective district officials, as noted earlier.

Local leaders

The biggest capacity-building task lies at district level. The Environmental District Officers were the first targets in the training process for environmental management decentralisation. As coordinators and advisors at district level they were briefed every four months about any changes in the process. One of the challenges in the process of building capacity among these Officers was the fact that they had varying terms of reference that were dictated by conditions of their diverse donor support bases. Other challenges will be noted in relation to formal education.

Capacity-building of the rest of the District Assembly was another training priority, starting with the District Executive Committee. During the first two-week training workshop, the District Assembly was introduced to the legal and institutional framework of decentralised environmental management. During the work-away period, they formed the District Environment Sub-committee (DESC). The DESC was then trained in the establishment of institutions at sub-district level, and in local environmental analysis using environmental indicators to identify environmental "hotspots". They were equipped with skills in collecting and compiling data and reporting, functions which had been mostly unknown in the centralised system. During their work-away period they collected information on environmental issues, and conducted outreach programmes to form or strengthen Area Executive Committees, Village Natural Resource Committees, existing non-govermental organisations (NGOs) and other community-based groups. In the next training workshop they compiled a preliminary District *State of the Environment* Report. Before the fourth workshop there were national sector workshops to link the district and national environmental analyses and to cross-check the information, after which a national sector report was compiled. Finally, an implementation and monitoring workshop was held, which transformed the Environmental Action Plans into environmental micro-projects, and oriented the participants on how to tap funding.

The training process was new to most participants. Local governments and their capabilities do not figure prominently in national strategies for public service training, as officials at Area Development Committee and Village Development Committee levels are often not members of the national civil service.

Capacity-building at sub-district level presented perhaps the highest challenge, as members at that level are either traditional leaders, voluntary committee members or retired members of the community. The workshop training style was still followed, but continuity became a problem because every time a workshop was held, new persons attended. The training process was similar to the district-level training, but concentrated on building skills in environmental assessment and planning, in order to improve participation in projects. Participants were also oriented to fund-raising procedures.

Formal education

Since most Environmental District Officers are drawn from professions other than environmental management, they have been gradually orientated to environmental management issues, environmental impact assessment, environmental monitoring and auditing. Officers without social science backgrounds also received training to understand the socio-economic and political dimensions of environmental issues and environmental management processes.

A programme is underway to upgrade most of the 27 Environmental Development Officers to a Master of Science degree with the University of Malawi. This has acted as an incentive for most of them to remain in the districts while they study. However, once trained and upgraded, some moved to other jobs and this has been a significant challenge in the capacity-development process. Most qualified personnel are reluctant to work at district level, because of the general absence of infrastructure.

The private sector, NGOs and para-statals have also been included in the capacity-building process through existing training courses. For instance, the decentralised environmental management process has taken advantage of an Environmental Education Course for Industries in Malawi, jointly supported by the Occupational Health Directorate and the Southern African Development Community (SADC) Regional Environmental Education Centre, and an Environmental Education certificate course offered by Rhodes University in South Africa.

Networking

Networking can be an important informal strategy for building capacity. This lay behind the establishment of the Malawi Chapter of the Environmental Education Association of Southern Africa (EEASA). The aim was to link environmental educators, environmentalist and general educators within Malawi, and with counterparts in the rest of the region. The Malawi Chapter initiated an e-mail discussion group among the interested stakeholders. The impact has not been very significant, however, as it fails to reach many people, particularly at the sub-district levels.

General challenges

A number of general challenges in the process of decentralising environmental management in Malawi further shape capacity-building needs at various levels.

While both the Environmental Management Act and the Local Government Act outline obligations for decentralised environmental management, they are in practice far from being in harmony, and conflicts arise in the implementation process. Although the environmental management institutions have advanced in the decentralisation process, many other line institutions and sectors are still far from embarking on decentralisation. They seem to struggle to figure out their roles in a decentralised system, and to maintain a grip on power. A lack of consensus among central departments has been a further key constraint. This is partly due to clashing sectoral policies, as noted above.

The capacity to meet the challenges of decentralisation may depend on the level of political will at the centre, and the willingness of central departments to surrender authority over important areas of social policy, as has happened in other countries (Phillips, 2001). In Malawi local government structures struggle to perform new functions and mainly continue to act as bureaucratic instruments of the centre, rather than as generators of alternative values, preferences and aspirations.

A key factor in this process has been the inability of local governments to increase their revenue significantly. Central governments keep the most buoyant sources of revenue for themselves and in Malawi local government lacks adequate capacity for revenue collection. Local government efforts are further hampered by bureaucracy at the centre, resulting in the inadequate release of allocated funds. The ensuing extension of central authority over local matters has been frustrating people's participation.

The holding of resources by most environmental sectors' central offices has been exacerbated by a lack of clarity surrounding the prioritisation of funds to be disbursed from the District Development Fund. There is often a big discrepancy between community preferences (as reflected by local government choices) and public policies. Local governments prefer to prioritise infrastructural projects like schools and health centres above environmental projects, because of the perceived greater long-term benefits. Hence, central environmental sectors fear that resources will at district level be misallocated to non-environmental projects.

Central government officials tend to see local leaders merely as communicators and solicitors of support for national policies, and neither as channels through which the conditions and needs of the local communities are articulated, nor as mobilisers of local resources for promoting environmental management from the bottom up. And indeed, equity and efficiency in utilising scarce resources are important issues that cannot be secured solely by local decision-making; conflicts of interests between and within different localities have already been evident in some districts.

The difficulties of decentralised environmental management in Malawi have been compounded by natural resources being allocated common-use tenure. Strengthening traditional leaders to be entrusted with the distribution of some natural resources has proved, in many instances, to be counterproductive. There has been reported dominance by local elites, and imbalances in the allocation of resources, with the poor being further marginalised. Participatory environmental management processes must have "downward accountability" (Murombedzi, 1991). The devolution of public powers to chiefs and other hereditary or customary leaders may lack such downward accountability, but so can projects, administrations, NGOs or other organisations with non-representative committees, as well as groups that temporarily mobilise through orchestrated participatory processes.

Conclusion

There are many challenges in this far-reaching socio-political process. It is clear that massive efforts are required to build the capacities necessary for carrying out decentralised environmental management, not only at various government levels, but also among a critical mass of the broader population who need to support, motivate for and participate in the process. Not only is the capacity-building task vast, it is also complex. In a resource-poor environment local officials and their partners need background knowledge and skills to survey and analyse the state of the local environment and to contribute to development plans that do not destroy the already pressured natural resource base. At the same time they have to deal with the considerable challenges of reconciling competing interests, ensuring that the benefits of environmental management are shared equitably among stakeholders, and report to diverse authorities who themselves are often in need of capacity-building in relation to the process.

The structures close to local communities are meant to drive the necessary educational processes, but most of their members themselves fail to conceptualise the nature of decentralised environmental management processes.

Nevertheless, Malawi has started a far-reaching education, training and awareness-raising programme to build local capacity. The process will need constant monitoring and review in order to determine its effectiveness, so that it can adequately fulfil its central role in the process of decentralising environmental management processes.

Acknowledgement

John Balarin introduced the author to Environmental Management Decentralisation. Designing the training processes and manual with Luke Malembo and Alifeyo Banda, and the discussions that accompanied the Capacity Development in Environment Project, were memorable experiences. Paulo Chiziwa was always at hand. The Project tested processes in the Karonga DEAP, Lake Chilwa Wetland Management, and Lakeshore and District Environmental Management Projects. All were Malawi Government and DANIDA-funded projects.

C H A P T E R   F I V E

Policy Playing out in the Field:
A Case Study of the Implementation of Sustainable Agriculture in Uganda

Daniel Babikwa, Uganda

Introduction

This chapter examines the implementation of environmental policies originating from the United Nations Conference on Environment and Development, ten years ago. It describes features of Uganda's environmental policy, specifically key assumptions on process and anticipated outcomes and the way policy has played out in the field, with emerging successes and challenges. It is based on a case study of Volunteer Efforts for Development Concerns (VEDCO), a rural NGO trying to translate Uganda's environmental policy into action at community level. VEDCO is committed to fostering sustainable development and poverty alleviation through capacity-building in sustainable agriculture, micro-finance, agricultural trade and marketing. The research, conducted in the rural district of Luwero in central Uganda, entailed 18 months' full-time engagement in the field in a participatory action research process with the NGO. The case illustrates some of the ways in which NGOs in Uganda have grappled with the implementation of the national environmental policy. It reveals contradictions in national policies, but also contradictory trends in the NGO, as well as the dynamic contextual factors that shape the directions of implementation and outcomes.

Background to Uganda's environmental policy

By the end of the 1980s, the Ugandan government was becoming increasingly aware of the impact of environmental degradation on the economy and the need to respond was urgent. Available evidence showed an astounding annual loss of up to 12 per cent of Gross Domestic Product due to environmental degradation (Converey, 1992) through soil erosion; deforestation in the watersheds; inappropriate farming methods; draining of wetlands for agricultural purposes in the densely populated areas; overgrazing; and the over-exploitation of lakes and rivers (Ministry of Natural Resources [MNR], 1995).

In 1992, Uganda participated in the United Nations Conference on Environment and Development (UNCED) in Rio de Janeiro and committed itself to environmental protection by signing the various provisions, declarations and statements of principle. In 1994 the country published its National Environmental Policy in line with the objectives of Agenda 21. The policy aims at ensuring

> [s]ustainable social and economic development which maintains or enhances environmental quality and resource productivity on a long-term basis that meets the needs of the present generations without compromising the ability of the future generations to meet their own (MNR, 1995).

The policy outlines six objectives which all emphasise social, economic and environmental sustainability as the core of the development process, a holistic approach to environmental management and the importance of community participation in sustainable resource utilisation:

- enhance health and quality of life of all Ugandans and promote long-term sustainable economic development through sound environmental- and natural-resource management and use;
- integrate environmental concerns in all development-oriented policies, planning and activities at national, district and local levels, with participation of the people;
- conserve, preserve and restore ecosystems and maintain ecological processes and life-support systems including conservation of national biological diversity;
- optimise resource use and achieve a sustainable level of resource consumption;
- raise public awareness to understand and appreciate linkages between environment and development;
- ensure individual and community participation in environmental improvement activities.

The policy was informed and guided by environmental principles adopted at the Rio Declaration (MNR, 1995; Quarrie, 1992), which include that:

- the development of Uganda should be based on sustainable natural resource use and sound management;
- security of land and resource tenure is a fundamental requirement of sustainable natural resource management;
- long-term food security depends on sustainable natural resource and environmental management;
- environmentally friendly, socially acceptable and affordable technologies should be developed and disseminated for effective use of natural resources;
- conditions and opportunities for communities and individual resource managers to manage their own natural resources and the environment sustainably should be created and facilitated.

Among the key implementation strategies of the national environmental policy are "public participation" and "international cooperation" (MNR, 1995).

The Environmental Investment Programme was set up to implement these strategies. The purpose of the programme was to ensure "a balance between socio-economic growth and sustainable use of natural and environmental resources" (MNR, 1995). The programme identified capacity-building in environmental management, enhancement of resource productivity, conservation and use of biodiversity resources, environmental education and public awareness as the central areas of investment to ensure sustainable development.

The National Environmental Management Authority (NEMA) was established in 1995 to oversee, coordinate, guide and support the implementation of the policy and its related activities. The implementers of the policy include government ministries, local authorities, research and academic institutions, the private sector and NGOs.

VEDCO: the organisation and its programmes

Community-based NGOs are seen as important initiators of local projects and potential resource organisations, capable of offering specialised advice to NEMA. Their particular importance lies in their potential to act as intermediaries linking grassroots activities to broader activities and policies at district and national levels, while at the same time working as convenient and effective agencies for administering community-targeted incentives (MNR, 1995), and as facilitators and mobilisers for environmental education and training in support of government programmes.

For these reasons, and because of my interest in community-based and participatory development work, I decided to focus my research on the translation of policy into action in a community-based NGO programme. In principle, community-based programmes are relevant and appropriate as they are developed to be responsive to the concerns of the local people. VEDCO also appeared to have holistic and integrated programmes with qualified staff as well as a management team aware of the integral nature of the environment and development process.

VEDCO's vision is as follows:

> Improved quality of life for small-holder farmers and entrepreneurs through sustainable economic empowerment demonstrable in the communities' sustainable utilisation of available resources, ability to negotiate for support services in the areas of food security, marketing, and increase in household incomes (people should be empowered to independently mobilise and utilise resources and services to meet their needs in life) (VEDCO, 1999).

The NGO's mission is to achieve this vision through capacity-building in sustainable agriculture, micro-financing and agri-business development.

VEDCO was founded in 1986 as an association of university students whose parents had been cut off in the Luwero Triangle, an active war zone during the guerrilla campaign that brought the current government to power. The aim was to trace the whereabouts of family members. After the war it took on a new role of responding to the challenges of poverty, displacement and devastation. It operated as a self-help organisation in the district to distribute aid to the resettling returnees (VEDCO, 2000).

participatory action research encourages, we were able to identify the main problems and challenges encountered by the programme and to implement significant adjustments to the structure, focus and methodology.

In the eighth month, we began involving farmers in the generation of information about the programme and in making decisions about the future. Through our engagement with the farmers in the research processes we discovered that the programme focused on what most farmers regarded as secondary issues. It was also realised that the baseline study on which the current programme was based had erred in assuming that all participants had identical interests in farming.

Based on these insights, VEDCO decided to redo the baseline study in 2000, with strict attention to the principles of participatory development stipulated in the NGO's policy. The outcomes of the new baseline study led to the redrawing of most of the strategies and content areas of the programme. For instance, general workshops for practical skills in agriculture were scrapped in favour of more specific training, where farmers were grouped according to interest and offered training that met their particular interests and needs. Exposure visits to successful Sustainable Agriculture farmers were introduced instead of visits to agricultural research stations where the resources were beyond the imagination of the richest farmer in the community. With these and many other changes, farmer behaviour changed tremendously. Absconding from workshops and non-implementation of the new farming practices disappeared from the extension workers' list of problems. It was now also possible for the organisation to address capacity gaps among staff, to enable them to deliver quality services to farmers.

The most interesting outcome in my view was the change in the dependency spirit among farmers. While they had previously demanded handouts including transport and lunch during training workshops, they now willingly paid their own transport fares and provided their own meals.

Training in participatory methods equipped the extension workers with essential skills in facilitation and participatory training, and their approach to training changed to a more dynamic, interactive one. Both the extension workers and the farmers acknowledged this transformation, as observed in various comments, including the following:

> The way they conduct training these days is good because in the past they just used to call us to attend training without finding out what we were interested in. One time they called us to attend training at Katikamu on rabbit rearing but we were very disappointed because even us who were interested in crop husbandry found ourselves wasting our time listening to things we were not interested in. The aspect I have come to like about the current approach is the way they involve us in choosing what to learn and following us up in our homes to give us support. In the past, different organisations like UNAFA and NARO called us for training in Luwero Community Centre but these people never came to see our gardens, so they never knew whether we implemented what they taught us. One wonders whether they were even interested at all in knowing whether we would implement what they taught or not.[2]

> One of the most obvious yet important changes I have experienced is the way I approach people in the community. My respect for farmers has increased; my patience and willingness to listen to their problems has also increased. In the past my main concern was how many farmers I visited, but these days I am more concerned with the quality of assistance I give to the farmer. I have also learnt to generate information from farmers and even to assess my own performance. I can tell when the farmers are not satisfied with my method of work and adjust; I also feel confident to meet people regardless of the numbers. I no longer fear challenge from farmers. I can accept them and use them as sources of learning for the group and myself. The other thing I learnt is to look at myself as a learner, not a teacher; to accept that I can be inadequate in some of the areas. Before, I thought I was supposed to have answers to all questions and whenever I failed I felt very low.[3]

The farmers' positive responses to the participatory approach to training confirmed that they only valued training that addressed their needs directly, that they valued being able to share their existing knowledge in the training sessions, that it was important to them to have this acknowledged as valuable, and to be free to challenge the trainers. This does not imply that all farmers were won over to become participative at all times, though. There were times when long-established habits of expecting the extension workers to arrive with ready-made learning packages reigned, and this frustrated the facilitators.

Learning moments: issues and challenges in implementing policy in a community context

The positive interview responses reflected above were collected towards the end of my research with VEDCO. To arrive at that point involved a challenging journey during which we had experienced great learning moments. The extension workers, their management and I were on a number of occasions confronted with challenges related to the community, the NGO and its funders. In this next section I note some of the contextual issues that influenced the implementation of a Sustainable Agriculture and Food Security programme by VEDCO. While some of them are rather negative, it is important to realise that they constituted a very important part of the learning process which enabled us to progress.

Poverty, land, gender

Poverty remains one of the major obstacles to effective implementation of the environmental policy at community level in Uganda. The farmers participating in this study equated poverty with the lack of money to meet basic needs for survival, inadequate access to resources, tools and implements for production, and with the inability to overcome such forces independently. Some farmers could not even afford containers or animal waste for preparing organic manure. Ironically, poverty made it almost impossible for some who participated in the programme to implement what they had learnt. As one farmer put it:

> People are poor and most of them depend on personal labour. It becomes difficult for many of us to practise the new skills and knowledge because some of them need money to be implemented. At times those people wait to see how we who implement benefit from the things we do, but unfortunately many of us are not good examples. I ... for instance grew eggplant in large numbers; when they matured I had nobody to sell to, yet I had wanted to use money from the sales to buy the necessary inputs. As a result my problem became worse. I wanted pesticides to kill the pests because the local concoctions had failed to kill all of them, but here I was, unable to complete the work I had embarked on. My situation became a point of reference for those who denigrated our involvement in the training. I could not attract those slow adopters because there was nothing good to learn from me. Actually, they laugh at me because some are better off than me.[4]

Another farmer lamented similarly and I could clearly see his plight, which, I believe, was shared by many of his compatriots:

> The training approach and content has been very good because it focused on the real things we have been trying to understand and now I can say that I know what I am doing, but there is one thing: Omwavu ne bwasoma taba n'amagezi ["It does not matter how much you educate a poor person, he will always remain a fool"]. Due to poverty we cannot implement, that is, get those materials like banana suckers, improved seeds. That is why we are still seen as fools; our knowledge is not reflected. Do you think a visitor will see me and believe I have any knowledge and wisdom? No, no. Because of poverty I am sitting on them.[5]

Landlessness did not only compound the problem of poverty, it was also a major obstacle to the implementation of sustainable agricultural practices. Whilst the average plot of land, at three acres per household (Kyaddondo & Kyomuhendo, 2000), looks like a fair distribution in the district, the production dynamics at household level suggest the opposite. In most households, more than half of the land is under perennial crops like coffee and banana. Land used for such crops cannot be used for any other productive purposes, and yet these crops, coffee in particular, give yields only twice a year, while the farmer incurs maintenance costs all year round.

The situation is worse for tenant farmers. Without any written contracts or agreements, their fate is dependent on the mercy of the landowner. There is widespread exploitation, as when, for example, landowners claim back land lent or hired out just as soon as the tenant farmer has improved it. The landless were thus reluctant to improve the condition of fields through hard work, knowing that owners would claim it sooner rather than later.

While women were the main tillers of the land and frequently participated in the VEDCO programme, in most of the cases they did not own the land, or did not have full powers to use it and to implement the sustainable agricultural practices they had learnt. It often took a lot of negotiation for a female farmer to begin using land freely, and in most cases such freedom occurred only when she was to use the land for household food production, or for a joint project with her husband, in which case she would be the lesser partner. During interviews I conducted, most women said that they had no land of their own. While this was obvious with married women, it sounded strange coming from widows. One would have expected them to become the owners after the death of their husbands, but instead, they would say, "the land belongs to my late husband". Such was the extent of women's vulnerability in matters of land ownership.

There were two women who had actively participated in all training and had excelled in the implementation of sustainable agricultural practices, to the extent that VEDCO took them on as lead farmers to train others. Unfortunately, neither of their husbands welcomed this. One of the women, a second wife in a polygamous household, was abandoned by her husband, who claimed that she could take care of her own needs and those of her children, since she was now also making money. The second was prevented from selling any crop grown on her husband's land and when she borrowed land to continue farming, she was divorced for disobedience. This created a sense of insecurity that discouraged many women from making full use of the knowledge and skills acquired.

Tensions and emerging contradictions

Bringing the three concepts of economy, society and environment together under the umbrella of sustainability is a good principle. But the concept of sustainable development itself is faced with the major contradiction of having to exist in a global capitalistic system which, by its very nature, is rooted in the exploitation of natural and human resources. In Uganda, this contradiction is evident in various national policies. Outside NEMA, the official development discourse clearly reflects neo-classical developmentalist[6] thinking (Mshana, 1992) where concepts such as the modernisation of agriculture are used uncritically, and mechanisation and monoculture are seen as strategies for modernising the sector. Low agricultural productivity is blamed on the non-use of chemical fertilisers and pesticides,[7] with little recognition of the environmentally friendly alternatives being explored by groups like VEDCO.

VEDCO also contains many contradictions. These relate to the organisation's changes in priorities, focus and character. It has come a long way from its origins as a social welfare foundation, with goals and interests geared towards addressing social conditions in the aftermath of war, to its current emphasis on corporate business principles. The focus on economic goals has affected the interpretation of other goals. Farmers were originally referred to as partners, but are currently called clients or customers by management.

Pressurised by donors' demands for accountability and delivery (discussed in greater detail below), VEDCO has struggled to perform. As one farmer observed:

> The training workshops especially on sustainable agriculture were good at the theoretical level, but the practical part was not done well in this village. They came in the middle of the farming season and tried to work very hard to catch the season, by making people work non-stop. Some thought it would always be like that *Kasiribaggo* [marathon]. Many people failed to cope with that speed and decided that they would not be able to manage VEDCO's method of work and pulled out of the entire activity. The elderly and the youth in particular did not approve of the approach. VEDCO wanted the change to occur very abruptly; the youth saw the use of demonstration as an approach as time-consuming, and not likely to reward quickly, so they pulled out.

This farmer's response refers to several other tensions in the situation on the ground. One of them is the "quick-fix" mentality dominant among youths in the villages where VEDCO worked. The nature of sustainable agriculture – being slow in yielding results and requiring much physical labour in the process – was a major disincentive to these youths who were interested in earning cash quickly. As a result they mostly shunned the training and favoured environmentally damaging activities like charcoal-making, firewood-cutting, sand-mining and brick-baking. The cumbersome nature of some of the sustainable agricultural practices was also a constraint among the older participants and women without access to assistance. They were frustrated by the amount of energy required for some of the sustainable agriculture practices. One woman told me that she had tried all she could, but her efforts to prevent soil erosion were being frustrated by the absence of her husband who had promised to dig the trenches she could not manage herself.

Another key tension noted by this farmer and others was the wish to use chemical pesticides. VEDCO discouraged this practice, preferring the use of organic fertilisers and environmentally-friendly organic pest-management concoctions using chillies, etc., but there were times when extension workers felt compelled to advise farmers to supplement with chemicals to control pests that had the potential to wipe out entire crops and severely affect food security and household income.

The paradox of participation amidst autocracy

As the above starts to illustrate, the extension workers were most troubled by the tensions and contradictions characterising the policy implementation process.

They found themselves at the bottom of a hierarchy. Despite the intention to run the organisation on participatory principles, management often acted autocratically. Many decisions were imposed on the programme officers and extension workers who were then compelled to do the same at community level. Such actions led to a feeling of disempowerment among all subordinate staff in the organisation.

While the autocratic tendencies were real, it was interesting to discover that there was no callous brute in the saga; the circumstances in which individuals and the organisation as a whole operated seemed to shape the conditions for autocracy. The NGO executive was a victim to the demands of the donors. This does not suggest that the donors were in a much better position. Just as the NGO had to struggle to impress donors in order to get the next financial release, the donors themselves were labouring to justify their relevance, both locally and internationally. There was an illuminating scenario during a strategic review and planning workshop for the organisation, when programme officers and extension workers complained about the despotic tone of the communication from the coordinators, often coming in the form of: "the Director has said this must be done". These directives often interfered with planned activities at a time when no meaningful contingency plans could be put in place. In his defence the Director noted:

> I know this has happened, but not very often – one time two donors were coming, they gave an abrupt notice and I thought we were not ready for them, and I rung the programme officer and told him, these people are here, what do we do? Is there anything for them to see or what do you suggest, do you think we can have them ... ?"

I also had the opportunity to participate in a joint NGO-donors meeting where the progress of programmes and future collaboration were discussed. This was a great learning moment for me regarding the thinking and attitude of some donors towards their local partners. I was at the same time able to realise the plight of the donors, themselves "beggars" from either home governments or other financial and development institutions. In relation to the nature of programme indicators, and the NGO's suggestion for more long-term, qualitative indicators of programme impact at community level, one donor representative kept asking: "How do we convince the tax-payers at home with such indicators? You know this is tax-payers' money!"

While the donor representative's plea for certain indicators of accountability seemed reasonable, it failed to take into account the realities of the context in which the funds were to be spent. It did not consider that VEDCO, for example, was working with people it did not control. The farmers (partners or clients) were volunteers with nothing forcefully binding them to the organisation or its development initiatives.

The dependency of the NGO on donor funds was a critically disempowering factor. The threat of losing funding cowed the NGO into obedience to unrealistic or inappropriate demands. The principle of accounting and reporting to the donor agency, so that it could in turn account to the source of funding, appeared sound in theory. But in practice, the principle relied on both subtle and not-so-subtle coercion. The NGO was literally bullied by threats of fund withdrawal or non-release in the following quarter. This pushed its management to put pressure on staff who in turn pressurised the farmers, who ended up undertaking activities not because they necessarily understood them, but simply because the NGO wanted them done.

There were several reports of community plans being altered or shelved without consultation, and of VEDCO meetings and activities either interrupted or cancelled because a donor was visiting and wanted to meet staff or have an activity report drawn up.

Flawed participation and assumptions about participation

Manipulating and abusing communities' goodwill, in agreeing to become partners with the NGO and its donor friends, is detrimental to the spirit of participation and collective problem-solving that the national policy framework and international guidelines quoted at the start of this chapter wish to foster. VEDCO had not fully succeeded in nurturing a spirit of participation. As earlier observed, there still existed autocratic tendencies which discouraged the growth of participatory training methods and approaches to implementation. The factors which restricted participation in VEDCO ranged from adherence to and expectations of old-style top-down training among extension workers and farmers, partly due to a lack of training in participatory methods; to the donor dependence and attempts to become financially sustainable which turned the organisation's focus away from participatory processes.

There were also dubious assumptions about participatory processes evident in the NGO's work. For example, in the "participatory" baseline study that informed VEDCO's original programme, "participatory" techniques were used merely as tools for extracting information, defeating the empowering goals associated with participatory methodology. Neither the communities that gave the information nor the frontline staff doing the final implementation at community level was involved in the final analysis of findings and design of the programme. In cases where some community representatives were invited to participate in the dissemination workshops, it was not necessarily those who had been part of the study. To the many people who participated in the study, there was no link whatsoever between the study in which they participated and the programme that was introduced as a result, for they had been left out during some of the most essential stages. It was thus not surprising that many of the outcomes of the baseline study failed to appeal to community members, and the study had to be repeated two years later.

Another questionable assumption of the programme was that after they had been trained by the NGO, a core of lead farmers would go to their respective communities, recruit villagers and replicate the same training. It was assumed that good farmers would be good trainers and that people were communal and willing to work together. Both assumptions were mistaken. Those lead farmers who did undertake training often did so in a technicist, demonstrative manner.[8] Also, farmers in most cases did not participate in these activities beyond the collective training and mandatory setting up of demonstration gardens, overseen by VEDCO staff. Not all lead farmers were willing to share knowledge with others. Some of them saw sharing as the invitation of competition, and so actively withheld information. These factors gravely militated against the anticipated participatory implementation strategy.

The assumption that communities are homogenous and ready to collaborate for a common good is an over-simplification of their nature and the dynamics that characterise them (see Katahoire & Ndidde, 1996:9). Similar as people may look in their poverty, great individual differences exist in outlook, interest and aspiration. The farmers that VEDCO worked with were evidently individualistic at heart, and communal and collective approaches only worked in those situations which centrally benefited the group, like training workshops. The culture of the communities of Luwero is for people to "look after themselves". Thus, farmers often saw working on a communal demonstration garden as giving free labour to the person hosting the demonstration; when asked to offer land for the demonstration they would often give that piece they had themselves failed to farm, either a mature bush, an area where tree stumps had to be uprooted, or one with soils compacted by livestock. Other dynamics that hampered the programme included the power equation at household level; power struggles in community leadership; nepotism in opportunity-allocation; divided loyalties between different NGOs; personal interests and competing demands.

These dynamics are a challenge to the philosophy underpinning the concept of participation and participatory community endeavours. While I do not dispute the value of participatory approaches to development, I am of the view that participatory endeavours should be contextually located and adapted. Participation is socio-economically, politically and culturally influenced, a fact often ignored or glossed over by development agencies. This leads to the imposition of a particular view of participation that is only aligned with the ideology and context of the agency.

Conclusion

Uganda's National Environmental Policy is sound as far as the theory of sustainable development goes, and is in line with the principles of Agenda 21. But how easy is it to implement this policy in the field?

The attempt by VEDCO at participatory community-based policy implementation reveals a number of useful insights that development practitioners, scholars and planners need to consider. It has been demonstrated that the use of participatory methods to implement policy is a complex affair. While it is easy to suggest a participatory approach, it can only succeed if one genuinely considers contextual matters around all the key stakeholders. Tensions arising from contradictions and conflicting agendas within and between national policies, local interests, the NGO and the donor can be major hindrances in policy implementation. Issues like dependency on external funding and a lack of consideration for each other's contexts can frustrate success. Hopefully, in the final analysis, this case illustrates that participatory policy implementation is complex, but not impossible. It is necessary for the implementers to demonstrate the values of self-criticism, reflexivity and collaboration. In particular, the significant shifts observed after the extension workers started to bring a more participatory approach to their work, guided by ongoing reflection on their own practice in a supportive participatory action research process, demonstrates the values of implementing policy in a self-critical, reflexive and collaborative manner.

Endnotes

1 Reijntjes & Waters-Buyer (1992) describe LEISA as a farming practice that seeks to optimise the use of locally available resources by maintaining the synergetic effects of different components of the farming system: plants, animals, soil, water, climate and people. External inputs are used only to provide deficient elements in the ecosystem and to enhance available biophysical and human resources. LEISA aims at stable and long-lasting production levels and site-specific farming developed within each ecological and socio-economic system. It incorporates the best components of local farmers' knowledge and practices, ecological agriculture developed elsewhere, and conventional scientific knowledge.

2 This was a response in an interview with one of the VEDCO farmers who had attended training before and after the training in participatory methods was implemented. The farmer had also participated in training by several other NGOs, which gave her a good opportunity to compare the different experiences.

3 This was the response of one of the extension workers when I asked him if he had experienced any changes in his way of working since he took part in the training on participatory methods.

4 This was the response of one of the active farmers who was feeling frustrated by poverty. I had asked him why many people were reluctant to implement sustainable agriculture practices learnt in the workshops.

5 This was a response from one of the farmers I interviewed who had participated in the various farmers' workshops in one of the communities.

6 Mshana (1992:17) defines developmentalism as the reduction of the concept of development to the mere production of wealth by use of science and technology accompanied by a view that it is universally applicable and necessary for societies.

7 The Plan for Modernisation of Agriculture (MAAIF & MFPED, 2000:75) cites Uganda's rate of using chemical fertilisers as the lowest in the world and suggests an "intensive fertiliser drive" at farm level.

8 Most of the NGO staff were amateurs in the area of participatory facilitation, being fresh from school and more comfortable with the traditional lecture mode of teaching. The more experienced field workers were also confident with their subject matter but not with the methodology. Hence, when they trained lead farmers to train fellow farmers in the community, they gave them technical knowledge in sustainable agriculture, but not in community training.

CHAPTER SIX

The Evolution of People-and-Parks Relationships

In South Africa's National Conservation Organisation

Kevin Moore and Lynette Masuku van Damme, South Africa

Introduction

This chapter relates aspects of the story of conservation in South Africa, paying particular attention to how South African National Parks (SANParks) has in recent years come to explore ways of building more positive relationships with neighbouring communities. The story is told by two social ecologists employed by SANParks. Touching on political ideologies and their influences on conservation from as early as 1811, the shared story illuminates how global trends and the local political climate influence conservationists to consider other ways of managing parks.

South Africa's conservation agencies have been highly regarded for their species conservation successes during the last century. In the people-and-parks realm, however, they have had far fewer success stories. National parks (like provincial nature reserves), have traditionally been managed as islands within the broader environmental landscape. Even problems directly caused by the establishment of a park (e.g. wild animals breaking through fences and causing damage on communal land), were not adequately addressed. Legal responsibility rested with the provincial conservation authorities, and not with park officials. This has been just one source of conflict and resentment between parks and neighbouring communities (Cock & Fig, 2002).

Conservation and development initiatives both within and around protected natural areas (PNAs) are greatly influenced by the broader political environment on local, regional, national and international levels. Indeed, conservation through the establishment of PNAs is a political matter, as it directly intervenes in the relationship between people and their environment (often between them and their means of production). The creation of PNAs in South Africa often failed to take this into account, resulting in conflict with displaced and neighbouring communities, and relationships mostly characterised by lasting resentment.

Influenced by national and international developments in the 1990s, SANParks introduced a Social Ecology programme in an attempt to improve the organisation's relationships with communities neighbouring parks. This chapter tells the story of these developments and associated challenges, and concludes with reference to some new obstacles facing the organisation at the start of the next environment-development epoch.

Some features of pre-1900 conservation in South Africa

The roots of South Africa's national conservation organisation can be traced back to the height of colonial rule in southern Africa. Its mandate and in fact the very reason for its establishment was greatly influenced by the trends of the early 1900s. A brief look at the pre-1900 history of conservation in Africa will provide a sense of how this would later influence the way in which conservation would be practised and enforced.

There is considerable evidence that at the time of colonial settlement, there were long-established and successful ways through which "Africans had ensured their own survival and that of the soils, plants and creatures which they needed in order to live and which form a basic part of the texture and meaning of rural, non-industrial existence" (Anderson & Grove, 1993:1).

European conservation initiatives failed to recognise the value of indigenous people's understanding of wildlife and land management. This aspect of colonial conservation, which stretched into the present, has been discussed elsewhere (e.g. Carruthers, 1995; O'Donoghue, 1998; see also O'Donoghue & Neluvhalani, this volume).

The next feature to note is that colonial conservation interests were clearly socio-economic. An early example from the Cape, which is regarded as the origin of European conservation interest in the subcontinent, is the reservation by the Royal Navy in 1811 of the use of forests in Plettenberg Bay in the southern Cape. This reservation was withdrawn when it was realised that the timber was not suitable for boat building (Anderson & Grove, 1993).

Even at this early stage politics and the power of the electorate would influence the way in which the environment was to be conserved. In 1858 Dr Ludwig Pappe, an Austrian surgeon, was appointed Chief Botanist of the Cape Colony. His main tasks were to develop a programme for soil, water, forest and species conservation (generally in that order of priority) in the Knysna and Tsitsikamma forests. Pappe's success with the politicians of the day was due to his adoption of the argument that forest destruction influenced rainfall. His conservation aim of species preservation was therefore achieved by highlighting the human suffering which would be caused by drought and thus the indiscriminate utilisation of the southern Cape forests. Droughts in 1847 and 1862 helped to swing public opinion and culminated in government setting up what were effectively the first state-run game reserves in Africa, in the Knysna and Tsitsikamma forests (Anderson & Grove, 1993). However, as soon as scientists could no longer make the drought threat credible, the temporary support of the settler community came to an end and state protection of the forests collapsed. European settlers, whose land- and capital-intensive activities were dependent on the continued deforestation of the environment, hastened this collapse. It was not until the 1930s, when the interest of peasant agriculture became politically expedient, that botanical experts began exerting their influence on land use.

It was the over-exploitation of wildlife through indiscriminate hunting which was to have a greater influence on conservation, eventually leading to the establishment of PNAs. Historically there were three principal phases in European hunting in the region. First, there was commercial hunting which concentrated on ivory and skins; then came the expansion of the colonies when game meat was a crucial protein supplement of the settlers. The third phase saw the transformation of hunting into the hunt, the objectives of which were the collection of trophies and natural history specimens, the establishment of records, and the pursuit of manliness and moral edification through sportsmanship (MacKenzie, 1993).

The first two phases were the periods of massive destruction of African game by the settlers' guns. Prior to these, traditional hunting methods had far less impact on the environment. The wanton destruction of wildlife that followed the acquisition of modern guns totally decimated once teeming herds.[1] The third phase of hunting saw wildlife being slaughtered for Europeans' sport. Indigenous hunters' subsistence use of wildlife was referred to as poaching and dealt with mercilessly by those who regarded blood sport as noble.[2]

The Cape Act for the Preservation of Game of 1886 was the first adequate legislation to be introduced for wildlife conservation. The resulting regulations and laws were, however, implemented to restrict hunting to sportsmen and further excluded Africans from wildlife utilisation.

The establishment of protected natural areas

To address the demise of game, a number of game protection associations were established, with membership limited to affluent gentlemen (Pringle, 1983). Carruthers (1995) describes how these early wildlife protection bodies continued the trend of protecting hunting as a sport and preventing Africans from subsistence hunting. It was at this point in South Africa's history that systems of protected natural areas were established.

South Africa followed the example of the US federal government which established Yellowstone, the world's first "National Park … to be preserved and managed … for the enjoyment and benefit of the people" in 1872 (Sellars, 1997:7). Early efforts eventually culminated in the establishment of the Kruger National Park and National Parks Board in 1926. From the start, and for much of their history, enjoyment and benefit of the Parks in both the USA and South Africa, according to Dowie (1996, cited in Khan, 2002), was limited to upper- and middle-class whites. The establishment of PNAs in South Africa went hand in hand with the forceful eviction of indigenous people (Carruthers, 1995). The conservation history of acting out of self-interest, on economic, political and increasingly racial grounds, continued. The ideal PNA was perceived as one which was uninhabited by people and which catered for mobile, affluent visitors. Indigenous people were excluded, except to serve in menial roles (Khan, 2002).

The practicalities of visiting a national park also denied many Africans access. Those who could afford to visit parks were seldom tolerated, despite the fact that the National Parks Act stated that the parks had been established for the benefit of the South African public (Khan, 2002). According to Carruthers (1995) blacks were not welcome, and the only provision for the separate accommodation of blacks was a (little-known) tented camp established in the Kruger National Park in 1932. In 1937 a few huts were erected in Skukuza for Indian South Africans. It was not only extremely difficult for Africans to secure accommodation in the Park, but they were also made to feel unwelcome through being cast in the racist stereotype of the poacher (Carruthers, 1995).

The National Party's election victory in 1948 marked the beginning of racial separatism for the next 46 years. This included what Khan (2002) calls the extreme politicisation of environmental conservation and the institutionalisation of environmental racism. Legislation passed during the apartheid era had major implications. The Group Areas Act (RSA, 1950) and the Separate Amenities Act (RSA, 1953) prevented Africans from visiting parks. The latter Act stipulated that public areas could be reserved for a particular race and that this action could not be ruled invalid on the grounds that the separate facilities were not substantially equal (Horrell, 1968). The Act was used by state departments to provide a grossly unequal distribution of recreational amenities, including those in PNAs (Hugo, 1974; Khan, 1990). This exclusion undoubtedly had a detrimental effect on attitudes and perceptions towards conservation efforts, resulting in hostility and a lack of interest, and contributing to the stunting of the environmental movement, restricting it to a narrow white support base (Khan, 2002).

The national parks were managed by a statutory board whose members were exclusively appointed from the ranks of white males who were generally closely aligned with Afrikaner nationalism (Khan, 2002). The organisation as a whole was controlled by white Afrikaner males and largely reflected the culture and practice of apartheid (Carruthers, 1995; Cock & Fig, 2002). Biodiversity preservation was the exclusive concern of conservationists in the organisation, whilst human and social issues were mostly ignored. "The notion of pristine wilderness and human exclusion, as exemplified by colonial attitudes, was sectional, and exacerbated national divisions along racial lines" (Cock & Fig, 2002:133).

The American military style of national park management, reflected in uniforms as well as in policies (Sellars, 1997), was mimicked by South African conservation agencies. Rangers and park managers were often ex-military men and several parks were used for military purposes.[3] This management style did little to encourage good relations with neighbours.

Early people-park interactions: environmental education and interpretation

The International Union for the Conservation of Nature and Natural Resources (IUCN) was among the first of the international conservation organisations to realise that the management of natural resources was not possible without making the links between conservation and development concerns. This led to the concept of sustainable utilisation proposed in the *World Conservation Strategy* (WCS) which the IUCN published in 1980 (with UNEP and WWF). The WCS showed how conservation can contribute to the development objectives of governments and suggested the sustainable utilisation of species and ecosystems as a key objective.

Arguing that conservation was both a logically useful and morally right thing to do, the WCS supported the need for PNAs, and in particular their role in maintaining ecological processes. The WCS also prompted governments to educate their people so that they themselves could manage their impact on the environment, thus giving prominence to the role of environmental education in conservation.

The recommendations of the WCS and its combination of utilitarian, technocentric and ecocentric arguments for conservation (Janse van Rensburg, 2001) were soon reflected in the management of South Africa's national parks. Interpretation officers (also called information officers, communication officers and trails rangers) were appointed in parks. These officers were primarily Afrikaans-speaking men with qualifications (usually post-graduate) in the natural sciences. The Kruger trails, based on the successful wilderness trails started by Ian Player in KwaZulu-Natal in the 1970s, became extremely popular. Trail rangers and their black trackers took groups of up to eight clients (mainly affluent whites) into the bush. The environmental education message encouraged living in harmony with nature through a change in environmental ethics. The Kruger trails became a financial success, with income far exceeding expenses in an example of how environmental education could (under certain circumstances) be financially self-sufficient.[4]

By 1990 interpretation officers had been appointed in nine national parks and regional offices, and three environmental education centres had been established. The tasks of the interpretation officers included environmental education, visitor interpretation, public relations, marketing and promotions. Visitor interpretation usually included static displays, talks and video or slide presentations; school visits were usually day trips on which a number of interpretation techniques were used. Like the American model, efforts concentrated on those holidaymakers, schools and youth groups who had the means to visit parks. While the number of tourists and school children who were accommodated was impressive – a reported 30 000 people per annum (Erasmus, 2002[5]) – these audiences were usually exclusively white. It was only in exceptional cases that educational work reached black communities. When this did happen, it took the form of camps for peer-leader groups.

The post-1994 period of transformation in South Africa changed the priorities of the organisation. At the time this did not include interpretation and the next few years would see a whittling down of the service to visitors and schools, who ironically would now also include black South Africans. The interpretation officers of the passing era either left the organisation or took on other functions. The largest and most successful interpretation centre in Skukuza was closed down.[6] A new beginning was to be made in SANParks, which would see the introduction of a more inclusive form of conservation aligned with South Africa's new democracy. Interpretation would become incorporated[7] into the environmental education function associated with the new approach to conservation, which SANParks called Social Ecology.

The transformation of a conservation organisation

In the 1990s local environmental NGOs did much to broaden environmental awareness, echoing international calls to link environment and development issues and for participatory approaches to both conservation and development.[8] The work of these and other groups was strongly aligned with South Africa's socio-political change towards democracy and they put considerable pressure on conservation agencies to change the way in which natural resources were managed. Under such pressure, and following the country's first democratic elections, the need to transform the leading conservation agency became a priority. In 1995 the first gender and racially representative board was elected. The organisation was renamed South African National Parks (SANParks) in 1996[9] and slowly but inexorably it began to address the problem of the exclusion and alienation of black South Africans.

The Constitution of the Republic of South Africa laid the basis for a new approach to conservation in SANParks. The new board adopted a new vision – "National parks will be the pride and joy of all South Africans" – and the following mission to achieve the vision: "To acquire and manage a system of national parks that represent the indigenous wildlife, vegetation, landscapes and associated cultural assets of South Africa for the pride and benefit of the nation" (SANParks, 1995).

The new mission and its interpretation in the Corporate Plan (SANParks, 1995) showed a fundamental shift from one-sided conservation practices to a more holistic, integrated natural and cultural heritage management. The new approach, which was to become embodied in the notion of Social Ecology, recognised ecological, cultural and socio-economic issues as all critical to the sound management of national parks. As the director of the newly-established Social Ecology Department and the first woman to be appointed to a senior position in the organisation put it:

> [t]his new thinking embraces the notion that the future of National [sic] parks is intertwined with the local communities living adjacent to them. Thus, National parks have

> to be integrated into the surrounding local communities for the future survival of protected areas (Dladla, 1998:foreword).

SANParks thus became committed to promoting a different concept of conservation, one linked to issues of development and human needs. Cock & Fig (2002) describe it as a concept that implies a harmonious relationship between people and parks and one which builds on traditional concepts of wilderness and wildlife in African indigenous cultures (see also SANParks, 1998).

The board also noted that under the new Constitution's provisions for land restitution, a number of legally recognised land claims could have a significant impact on national parks. The board therefore established a land claims committee and developed a policy which committed the organisation to settling claims through negotiation, mindful of the need to balance social justice with its mandate to acquire and manage land to conserve biodiversity (SANParks, 1998). A number of claims affecting national parks had been lodged with the Ministry of Land Affairs. Two of these, Makuleke-Kruger and Khomani San-Kgalagadi, were successfully resolved. In both cases the ownership of the land was returned to the claimants. However, the land use remained conservation and contractual parks were established with joint management between the community and SANParks. The Makuleke agreement was described by Koch (1998) as "a unique attempt to harmonise the protection of biological diversity with the interests of rural people".

Social Ecology as a core function within SANParks

SANParks's Corporate Plan (1998) describes Social Ecology as

> a strategy and process that conveys the philosophy and approach of SANParks to neighbouring communities and establishes mutually beneficial dialogues and partnerships with these communities. The process ensures that the views of the community are taken into account to the largest possible extent and are acted upon, that the park's existence is a direct benefit to neighbouring communities and that, in turn, communities adjacent to parks welcome the conservation efforts of the SANParks.

This definition was later modified to stipulate the inclusion of cultural values and resources in park development and management processes. Social Ecology was seen as an effort to bring about social justice, redress and development, and particularly to build more positive relationships with neighbouring communities. To these ends the following goals were adopted by the Social Ecology Department in 2000 (SANParks, 2001):

- establishing mutually beneficial partnerships with local stakeholders by taking the views and needs of local interests into account in park management;
- integrating cultural issues into the management of parks;
- promoting an appreciation of conservation among local stakeholders.

The programme comprises policy formulation and programme development and implementation, and is facilitated through a process that is interdisciplinary, participatory, community-orientated and educational (SANParks, 2001).

Four key performance areas[10] were identified in which both park- and corporate-based Social Ecologists would operate:

- **Environmental education**
  Programmes were created to enable both local communities and other South Africans to acquire, or rekindle, a knowledge and pride in natural and cultural heritage. The focus was primarily on the youth as the future custodians of the environment. Regional and national Environmental Youth Symposiums involved children from parks' neighbouring communities in action projects. The 1998 National Youth Symposium culminated in an Environmental Youth Charter, which was presented to the President of South Africa as the basis for establishing partnerships committed to social, cultural and economic actions aimed at promoting sustainable living.

- **Economic empowerment**
  The intention was that SANParks, as a resource-rich player in regional development, should provide economic opportunities for previously disadvantaged communities bordering its parks. The organisation adopted an economic empowerment policy, which aimed to establish "mutually beneficial partnerships" that were economically viable and sustainable. The facilitation of economic growth was indeed achieved through some pilot projects, for example in the Skukuza Alliance, which built facilities and provided business opportunities for local artists and traders. This project was initiated by the Kruger Social Ecology Unit, to promote authentic South African arts and crafts while ensuring that local artists around the Park derived economic benefits. Within five months, the project saw a collective increase in revenue for local artists from R2 000 per month to R50 000 per month. However, Social Ecology pilot projects for "community development"[11] took extensive effort to set up and sustain, were seldom successful on a sustainable basis, and always comprised but a drop in the bucket of rural and peri-urban poverty. Some in SANParks started questioning whether "economic empowerment" was indeed a function the conservation agency could fulfil. Others pointed out that in many cases "economic empowerment" initiatives amounted to little more than short-term charity which entrenched rather than resolved dependency.

- **Cultural heritage management**
  All parks have cultural heritage resources which need to be managed, conserved and interpreted.[12] Among them are non-tangible cultural resources such as the indigenous knowledge of tracking of the Kgalagadi San/Bushmen. Protecting and promoting cultural heritage is an important way in which national parks

can become the pride and joy of previously marginalised South Africans, through presenting a fuller picture of the country's distorted cultural history. As of 2002, SANParks will work with the South African Heritage Resource Agency to draw up cultural maps of all the parks, as part of and to inform park management plans and decision-making.

- **Liaison and partnership**
  Previously called Community Facilitation, this function strives to develop and nurture good relations with park neighbours by promoting their involvement, through advisory structures, in planning, management and the monitoring of issues relating to parks. By 2001 only a number of park committees had been established and of those, not all are functioning well. All parks, however, have other liaison and communication forums of numerous sorts covering a whole range of topics. SANParks is represented on these forums by the Park Managers, Social Ecologists and in some cases, Section Rangers.

The degree to which each of these four key performance areas was prioritised and implemented varied according to the specific park and community needs identified by park staff and local stakeholders, as well as park management's commitment.

The challenges of Social Ecology

As an approach to conservation in SANParks, Social Ecology has met a number of challenges during the past eight years.

When a Social Ecology General Manager was appointed by the Chief Executive to start up the process, it soon became evident that the organisation itself was in need of transformation. The Directorate agreed and the General Manager was tasked with heading up this process. As a result, Social Ecology itself became associated with the overall transformation of the organisation, which included staff and role changes. Many of the new Social Ecologists were the first black South Africans, and the first women, to be employed in skilled, managerial positions. It was therefore inevitable that they would be resisted as "affirmative-action appointments", and that the transformation process itself would be confused with the workings of Social Ecology. As in any long-established and previously largely homogenous organisation, change was viewed with scepticism, and many employees experienced the new order as a threat.

Another feature of the new Social Ecology appointments was that most were recruited from the social sciences, whereas previously, such appointments were only given to those with a natural science background. The introduction of specialists in the field of social science meant that a great deal of emphasis was placed on "community development", which became a key performance area of Social Ecologists at the time. The parameters and role of community development were not easily conceptualised by park employees and this period saw, for example, parks donating meat from culling programmes or building materials from demolished structures to neighbouring communities under the banner of community development.

A further challenge to Social Ecology was that under the National Parks Act of 1976, areas adjacent to parks were legally outside the jurisdiction of park officials. This meant that Social Ecologists were faced with a legal dilemma with regard to development work among local communities and outside parks.

But the greatest challenge of Social Ecology remained: what to do to contribute meaningfully to the socio-development of previously disadvantaged park neighbours. As it became clear that neither the newly appointed Social Ecologists with social science backgrounds, nor their colleagues with natural science backgrounds, were able to tackle these issues adequately, and that managers similarly lacked experience and role models to draw on, the organisation embarked on an extensive Social Ecology capacity-building programme between 1998 and 2001. The Danish financial and technical support which made this possible also included Social Ecology pilot projects. Through these initiatives the organisation started improving its Social Ecology approach to become more realistic and achievable. So, for example, the idea of community development was replaced with "community facilitation", which suggested that the conservation agency's Social Ecology staff could facilitate development (establish networks, call on more appropriate agencies, broker partnerships) but were not in a position to undertake development work themselves. Similarly, the idea of "economic empowerment" was made more realistic by opening up "economic opportunities" for members of previously disadvantaged communities. This included, for example, opportunities for new small-business contractors to assist with the clearing of alien vegetation in parks, in association with the Department of Water Affairs and Forestry's Working for Water initiatives.

The Social Ecology capacity-building initiative helped to plant the seeds of a learning culture in the organisation and Social Ecology helped SANParks to address its historically insular nature, characterised by a reticence to engage and seek critique. The emergence of a learning organisation was evident at a conference entitled *Towards Best Practice*, where SANParks hosted practitioners from a range of people-environment contexts. The event served as a platform where the role of Social Ecology could be deliberated in an open forum. It also shared the recommendations from a number of research projects commissioned by the organisation. An example provided was of a socio-economic baseline study of communities neighbouring the Addo Elephant National Park (SANParks, 2000b), which highlighted the need to broaden the understanding of mutually beneficial partnerships to include meaningful access to parks so that people would "benefit aesthetically, spiritually, culturally, educationally and otherwise..." (SANParks, 1999:19). The conference further gave voice – literally and figuratively – to several park neighbouring communities who took part in the proceedings (SANParks, Danced and GTZ, 2000).

The Danced capacity-building programme highlighted the extent to which the organisation's focus on environmental education and interpretation (including cultural interpretation) had been neglected while interactions with neighbours were limited to development and economic partnerships outside of parks. This led to the drafting of an Environmental Interpretation and Education Strategy in 2000, a status survey on environmental education and interpretation in 2001 and a related professional development programme for park staff, implemented in 2002.

To tackle the challenge of establishing mutually beneficial partnerships with neighbour communities, the Board had transformed Social Ecology into the status of a full Directorate in 1997. At its peak, this Directorate consisted of a Director and six Managers based at corporate level, with 50 park-based Social Ecologists on the ground (Moore, 2002). But this situation was not to last for long, as Social Ecology was about to experience yet another challenge – the need to show financial viability.

Financial sustainability – new challenges for SANParks and Social Ecology

The goals which SANParks took on and built into its Corporate Plan, compelled by international trends and the politics of the day, are not matched by the organisation's revenue. SANParks currently receives 17 per cent of its revenue from national government. The government grant has been pegged at R50 million annually for the past decade. In the 1999-to-2000 financial year, SANParks was only able to raise 66 per cent or R226 million of the rest of its revenue from own operations (*Financial Mail,* 2000). A steadily worsening financial situation forced the Chief Executive to launch Operation Prevail in March 2001 to keep the organisation from going into liquidation.

Operation Prevail included a commercialisation campaign, which involved the outsourcing of non-core functions such as shops, restaurants, cleaning services, security services and road maintenance. (This included, ironically, the outsourcing of financially successful ventures such as the wilderness trails in the Kruger National Park.) It having been noted that the human resources budget exceeded 60 per cent of the total budget, a staff restructuring programme was launched. This restructuring affected all departments, including Social Ecology.

The Corporate Unit of Social Ecology had already been restructured the previous year. The Director's position had been scaled down to that of General Manager and the six Management positions reduced to three. Under Operation Prevail the park-based Social Ecology staff was reduced from 50 to 34. Forty per cent of the staff at Geelbek Environmental Education Centre were retrenched. The conversion of the education and interpretation centre at Skukuza in the Kruger National Park to a tourism facility, and other decisions, may be linked to a view that education does not generate income, and is therefore not a priority within the new commercial model of managing parks. In a bid to become financially viable in an increasingly corporate or "private" framework, SANParks's educational and other services and facilities (e.g. night drives, trails) are outsourced to businesses. These partners need to charge competitive rates in order to be financially viable themselves. This privatisation-like process in SANParks, together with the necessity of growing tourism as a source of income, may combine to once more prevent the benefits of South Africa's national parks from being accessible "for all" – ironically, now that political restrictions have been removed.

The justification for Social Ecology staff cuts was that Social Ecology was to be incorporated into the core functions of the organisation, and for the responsibility of Social Ecology functions to be shared with other park employees such as Section Rangers. This assumes that Social Ecology functions are non-specialised and that most staff are able to undertake them. Experience shows that this is not the case. Also, while the plan to integrate Social Ecology into the core functioning of all departments is sound in principle, the process still needs to be driven by people, and preferably staff with specialist Social Ecology skills.

A recent survey showed that some parks (eight out of 20) now have no Social Ecologists and that others are too under-staffed to be effective in achieving the objectives of the Social Ecology policy (Moore, 2002). The decision to downscale the Social Ecology function could prove foolhardy in the long term. The cost of Social Ecology to the organisation in the 2000/2001 financial year amounted to a mere 1.9 per cent of the total expense budget. The cost to conservation of bad neighbour relations (resulting in poaching, theft, wilful over-harvesting, land claims, intimidation of tourists, poor publicity locally and internationally) have to our knowledge not yet been calculated, but is certainly considerable.

SANParks currently receives substantial governmental "poverty relief" funding for projects run in partnerships with the Department of Environmental Affairs and Tourism and the Department of Water Affairs and Forestry. An adequate staff complement is needed to implement these projects, without whom these funds cannot be used effectively, or used at all. SANParks continues to receive support from international donors, including the Global Environmental Fund, for the expansion of national parks, enabling the organisation to achieve its biodiversity obligations. Park expansions almost invariably involve the displacement of people and their families, the reduction of access to resources, and resultant ill-will. This underlines the need to continue paying attention to the realisation of the conservation approach that the organisation has called Social Ecology.

The spectre of NGOs creating bad press for conservation agencies and of communities invading the land that SANParks has been asked to protect may have receded. Yet it is doubtful whether the efforts of the past years, while considerable, were enough to counteract the legacy of decades not only of apartheid, but also of colonial conservation practices. Without adequate staff, skills and resources, the organisation may also struggle to deal with future people-park issues arising from new and expanding parks. It would seem that the next major challenge is to meet the organisation's transformational goals involving democracy, social justice, the promotion of cultural diversity, and the protection of natural and cultural heritage, with a financial and economic model that works with rather than against these goals.

Endnotes

1 In the 1860s two elephant hunters operating in the Zoutpansberg "bagged" an average of 700 elephants per year (Mackenzie, 1993). The tusks were exported to Europe and the bodies left to scavengers.
2 It is hard to overstate the case. The attitude was captured in many texts, e.g. *Sport in War* (1900), where the author, Baden-Powell, who was later to establish the largest youth organisation in the world, the Boy Scouts, described military operations as akin to sporting activities. This was particularly the case, he said, in the chase for "wild beasts of the human kind" which offered "plenty of excitement and novel experience" among the kopjes and broken country of Rhodesia (now Zimbabwe). Elsewhere he remarked that "the longest march seems short when one is hunting game ... lion, boar, buck, nigger or nothing" (Baden-Powell, 1897).
3 The Kruger National Park, for example, was used for the training of Defence Force soldiers, the covert supply of material to Renamo in Mozambique and the launch of chemical weapons attacks on Frelimo troops in 1992 (Cock, 1993).
4 Ironically, the trails were outsourced to a private company in 2002.
5 Gert Erasmus, Manager of Interpretation for Kruger National Park until the early 1990s. Personal communication, July 2002.
6 In 2001 the then dysfunctional Education and Interpretation Centre at Skukuza was turned into a tourist facility.
7 The interpretation function has recently gained greater importance as pressure on parks to become more financially self-sufficient gave rise to considerations on how to encourage higher visitor numbers through value-added experiences such as interpretation.
8 Political change was not the only motivation for transformational activities in SANParks. International trends in conservation played a role by emphasising the value of a parks- and people-approach to replace authoritarian preservationist approaches and to promote collaborative natural resource management with indigenous people, their identity, culture and needs (see Lupele in this volume). Universally, education was seen as a vital dimension of efforts to move towards sustainable development.
9 It included nine members chosen by a sub-committee of the national cabinet through a process of public nomination. A further nine members were nominated by the provincial premiers.
10 The introduction of a "performance-based" culture in the organisation, and references to what were previously "head office" staff and functions as "corporate", and to the SANParks's policy document as a Corporate Plan, reflect the start of another significant shift in the life of the national conservation agency.
11 Sponsored by the Danish government through Danced.
12 These include over 254 sites in the Kruger National Park, Eve's Footprint (dated to 117 000 BC) in the West Coast National Park, and San/Bushman rock art in Golden Gate Highlands National Park.

C H A P T E R   S E V E N

Industry and Sustainability:
A Re-View Through Critical Discourse Analysis

Leigh Price, Zimbabwe

Introduction

Within industry, the actions taken in response to an increasingly prominent environmental crisis include a number of possibilities. One of the most common responses is public environmental reporting. This chapter provides a perspective on industry's responses to calls for sustainable development, not through an overview of trends and initiatives in the region, but through an in-depth analysis of one instance of environmental reporting, namely the annual *Environmental Report 2000* of the Electricity Supply Commission of South Africa (ESKOM). This report was awarded the KPMG[1] Gold award for the best sustainability disclosure in an annual report in the Public Entities Category (*Annual Report 2000*) and the KPMG Gold award for the best Corporate Environmental Report in the South African Category (*Environmental Report 2000*). This report could thus be seen as an example of environmental "best practice" in industry in the region.

This chapter offers possible insights to those people in companies responsible for writing their environmental reports, to allow a greater understanding of the language and rhetoric they use. It also aims to help those who teach about the environment, by offering a possible approach for teachers that allows them to mediate between such texts as the ESKOM report and their students. Critical Discourse Analysis (CDA) of a document such as the ESKOM report can deepen students' understanding of the complexity of environmental discourses at play in society.

In the title of this chapter, the word "critical" is used to indicate an affiliation with CDA techniques. Nevertheless, it should be noted that this chapter uses only those CDA techniques which might easily be applied by the general public, in an attempt to demystify discourse analysis and open it up for more general use. This is not an ideologically "critical" chapter. While I use critical analysis techniques, I avoid positioning myself ideologically as either for or against ESKOM. I am not trying to uncover hidden foundational truths about ESKOM. Instead I try to show up contradictions and to problematise the use of language.

The title of the report used in the analysis is *Towards Sustainability*. The theme of sustainability will be emphasised in this analysis, since it is an important focus for the World Summit on Sustainable Development. It is also a word whose meaning and usefulness has been contested.

> The meaninglessness of such universal statements [as "sustainable development" and "sustainability"] ... and the arrogance of the white, upper-middle class, educated men who develop such statements, shines through. We need to be encouraging people to deconstruct these statements for the value they embody and the perspectives they contain (Gough, cited in Sauve, 1999:24, explanatory comment in brackets mine).

For example "sustainability" and the term from which it was derived, "sustainable development", do not indicate *what* is being sustained, or for the latter, *what* is to be developed. Given the frequency of the usage of the word "sustainability" it seems it may be useful to examine how this word is being used within the context of industry, and perhaps what the underlying perspective contained within it is.

Texts, such as the ESKOM environmental report, are instantiations of socially regulated discourses, and their processes of production and reception are socially constrained (Janks, 1997:329). Such a report is thus a form of social practice that is tied to a specific historical context and is a means by which existing social relations are reproduced or contested and particular interests are served. In this case, we are examining the social relations around environmental and sustainable development issues. We might therefore ask of the ESKOM environmental report questions that relate its discourse to the underlying environmental and development perspectives that the report embodies. Specifically, we might ask how these perspectives and their attendant relations of power are reproduced through the use of such words as "sustainability" within the report.

Methodology

In answering these questions, this chapter will adopt an approach that will include both engagement with, and estrangement from, the text. Engagement is necessary to allow the entry of the text into the confines of our subjectivity. It can also be described as "reading *with* the text". Here an important question would be: what does ESKOM management want to say to us in this report? However, engagement without estrangement is to submit to the power of the text regardless of one's own positions (Janks, 1997:330–331). In reading "with" the text, we are accepting the preferred reading and thus offering unquestioning support for the *status quo*. This chapter therefore uses CDA techniques to resist deliberately the text's apparent naturalness and tries to offer an alternative reading, i.e. it also attempts to "read *against* the text", by asking such questions as: how is the text positioned, and how is it positioning the reader? Whose interests are served by this positioning? Whose interests are negated? What are the consequences of this positioning? (Janks, 1997:329). This chapter will attempt to balance estrangement with engagement, as estrangement on its own is a refusal to allow otherness to enter, and may lead to an overly simple oppositionalising of, for example, ESKOM management versus the environmentalists of the world.

Reading "with" the text is a simple matter of cooperatively reading the text and taking it at face value. Reading "against" the text can also be a simple matter if one is very clearly starting from a position of estrangement (Janks, 1997:331). For example, the environmental report of a Zimbabwean chemical company claimed that its emissions of sulphur dioxide were well within the legal limits, implying (in the preferred reading) that its factory was safe. Yet when an asthmatic visited its premises, she immediately found it difficult to breathe. It was not hard for her to read against the preferred meaning of "legal limits", and decide that it certainly did not mean the factory was safe. She had found it easy to question the meaning of the term "legal limits" and whose interests it served.

However, where we are not immediately estranged from a text, it may be harder to question, and this is where a systematic Critical Discourse Analysis (CDA) approach can be helpful. In this chapter, I used a modification of Fairclough's CDA techniques, which involve the following: description (text analysis), interpretation (processing analysis) and explanation (social analysis). I focused primarily on just one text, the *ESKOM Environmental Report 2000*. As I was aware that this was a limited starting point, I also looked for the patterns I identified in this text in other related ESKOM texts. I asked the question: are my findings in the ESKOM environmental report able to explain quite small features of other related texts? This approach, as a way to validate CDA findings, is suggested in Fairclough (1992:238). In the interests of brevity, I have reported only on the parallel findings and deepening insights from one related text, namely the *ESKOM Annual Report 2001*.

In the interests of keeping the chapter to a reasonable length and in order to avoid an overly technical approach, I have not addressed such issues as modality, lexicalisation, nominalisation, or the use of the active or passive voice. Instead, this chapter focuses on the visual signs of the report, the thematic structure of the text and its information focus (especially any information omissions), to develop its main insights.

Description of the signs

The visual signs

The *ESKOM Environmental Report 2000* is printed on glossy paper and contains diagrams, graphs and colour pictures on each of its 56 pages. Running along the top of each page, heralding the text headings, is a pastiche made up of brightly coloured photographs of shiny fibre-optics, time-lapsed moving lights, electrical equipment and pylons. Several of the photos are taken from the perspective of the viewer looking upward, into the sky.

The cover of the report has three photographs. The central and largest one is a satellite picture of the world at night. Africa is in the foreground. It is mostly in darkness, with some lighting at its edges and in South Africa. The rest of the world, in comparison, appears to be very brightly lit. The second photograph is of electricity pylons and a coal-fired power plant. In the foreground is a field of grass and flowers; in the background are white fluffy clouds, which one could initially mistake for steam emissions from the power plant stacks. The third photograph is of a rural house, featuring a prominent solar power panel, and a homely scene of children playing and a mother sitting nearby, preparing food.

On page 5, a map of southern and central Africa is displayed over a full page. It is criss-crossed by red traces, indicating power lines, and scattered with multi-coloured dots, indicating power plants.

The verbal signs

The report provides a large amount of background information on ESKOM, its Environmental Management System (EMS), economic and social issues, and the performance of the company. It draws attention to the many sustainability projects in which ESKOM is involved and outlines the many mitigation measures underway to reduce impact on the environment. These measures focus on improved EMS implementation at certain coal sites, reduced water pollution and consumption, reduced air emissions, management of impacts on land and biodiversity through Environmental Management Programmes, land registers and pre-construction audits.

The report does not deal with how ESKOM disposes of its nuclear waste. It is nevertheless remarkable for its candour in reporting failures alongside successes. For example, it admits 21 legal contraventions, 66 environment-related complaints and reports, and, for the first time ever, gives details of the quantity and quality of emissions of its nuclear reactor. There are two sections of the report that deal directly with sustainability. The first is a section under the executive summary (p. 7) entitled *Towards Sustainability* and the second, given the same title, is an expanded version of the section in the executive summary, five pages long, at the very end of the report.

In the executive summary, the sustainability section is divided into two categories: demand-side options and supply-side options. There are two demand-side options: projects to improve industrial energy efficiency, and load management systems to reduce maximum demand at peak times, such as real-time pricing. There are six supply-side options:

- re-commissioning an old plant;
- researching and constructing new coal, gas and hydro plants;
- research into solar, wind, wave and water power;
- research into distributed generation technologies such as photovoltaic applications, fuel cell research and diesel for remote locations;
- importing and storing energy; and
- implementing a Pebble Bed Modular Reactor (PBMR) to generate nuclear power.

These options are expanded in the section towards the end of the report. The PBMR is mentioned twice, once very briefly in the executive summary (one sentence) and once in the later expanded version (four paragraphs).

Interpretation of the signs (processing analysis)

Reading with the text

The visual signs, with their bright lights and vision of the need in Africa for electricity, imply that ESKOM is a forward-thinking organisation, committed to the future and to providing Africa with energy. It plans to be in the forefront of energy production in the region. It takes the stance that a company can be both environmentally friendly and a competent producer (as is implied by the picture of the power station in the flower field) and that its activities will bring a better quality of life (implied in the picture of the rural family with solar energy) to all in southern Africa (evident in the map). ESKOM is a remarkable company in that it has a strong environmental management focus, a strong social conscience (for example, it sees itself as having an important role in the development of the region) and a strong commitment to being open about its environmental impacts. However, it is also a company run on shrewd economic lines that will confidently maintain its market superiority into the future. The demand for energy is so large that ESKOM is rapidly expanding production, through old and new technologies.

ESKOM has taken a pragmatic stance on the generation of energy using nuclear power. Given the need for energy in the developing region, directly related to its potential for development, and given the relatively clean nature of nuclear energy (apart from the radioactive waste), ESKOM considers nuclear reactors to be an important and viable option for sustainable power production in southern Africa. Thus it confidently places the PBMR in the sections on sustainability, as it will help secure future generations of power, at the same time reducing a variety of emissions associated with power generation.

Reading against the text

The image of Africa in darkness, with only South Africa and the rest of the world lit up, summons images of the colonialist ideal of enlightenment: bringing light into darkest Africa. That South Africa is, in this picture, also well-lit might also be read as an indication of South Africa's vision of itself as bringing "the light" to other African countries. Additionally, the map of Africa, showing ESKOM's commitment to bringing electricity to Africa, is strongly reminiscent of maps of Africa from Cecil John Rhodes's times: here, instead of electricity lines, there were railway lines. The "feel" or atmosphere around these maps was similar, the makers of both the ESKOM map and the Cape-to-Cairo railway maps proudly showing how Africa was going to be developed by forward-thinking visionaries intent on bringing enlightenment ideals into the Dark Continent. Although these pictures on their own could not sustain this reading, they could indicate a need to look out for evidence of neo-colonialist tendencies within the discourse embodied by the ESKOM environmental report, or other similar reports. Also, enlightenment in itself is perhaps only a problem if it is imposed, rather than being a mutual sharing of ideas and a mutual growth. Although the similarity of the ESKOM front-page pictures to colonial pictures is unnerving, perhaps a deeper reading of this and related texts is needed before conclusions are drawn.

The large amount of information provided and the status of the report, indicated by its high-gloss print, could be read as giving weight to the environment, but could also be read as a publicity stratagem (surely less gloss would be more environmentally friendly?). Does this report represent an attempt to obtain good publicity by making the most of the environmental focus of the time? One might hypothesise that if the environmental focus of the time were to change, then ESKOM's focus on the environment might change too.

Omissions

The absences in the ESKOM report may also be meaningful. To begin with, there is the relatively small coverage of the PBMR. This functions perhaps to make nuclear waste seem ordinary and thus worthy of no more importance than any other fuel option. The absence of any acknowledgement of the public debates and even outcries around the PBMR contributes to making it seem "natural" and logical. Should we be reading "with" the text on this one? Is nuclear power actually quite safe? The arguments for and against nuclear power are complicated, technical and full of uncertainty on both sides. The global trends in society would indicate a preference towards a precautionary approach to nuclear power because of its potential problems. For example, *Newsweek* reflects on the current attitude towards nuclear power as follows: "Nuclear: 'Future fuel' of the past 50 years, is now shrinking because of high costs and safety worries" (Sparks, 2002:42).

Nuclear waste seems to be the ultimate sustainability question because its negative consequences may be felt in such a distant future. Perhaps our approach to nuclear waste might be a good marker of how seriously we are taking sustainability to heart, where sustainability is given its original focus as "Development that meets the needs of the present *without compromising the ability of future generations* to meet their own needs" (WCED, 1987:43, my emphasis).

The absence of information on ESKOM's nuclear-waste disposal is an interesting omission given the candid flavour of the rest of report. This omission serves to make the other confessions seem possibly like a smokescreen or avoidance tactic. One has the impulse to ask what else has not been reported. If the company is at least seen to be trying to be open, is this enough? Bourdieu (1998) comments that often in society, it is enough to be seen to be trying to follow the rules, even if one is not actually, behind the scenes, following them. This tendency may be innocuous in some situations, but not where the welfare of people is at stake.

What does ESKOM mean by "sustainability"? Or, rather, whose interests are being served when ESKOM talks of "sustainability"? To answer these questions I refer to the report's introductory sentence on sustainability. In this sentence, ESKOM's approach to sustainability is stated as being a process which " ... provides the strategic framework for projections of supply-side and demand-side options that will need to be implemented to meet future energy demands" (2000:7).

There is no mention of the environment, and from this statement, it would appear that sustainability, for ESKOM, has more to do with business (energy production) sustainability than environmental sustainability. Thus one might argue that it is not the interests of the Earth (and thus sustainable living) that are being served, but those of the company.

To add to this, if one looks at the list of planned projects aimed at encouraging sustainability, most of the projects cannot be differentiated from projects designed to ensure the company's growth. For example, wind energy generation is starting to produce profits for companies in the USA and Europe (Earth Policy Institute, 2001). The projects that involve photovoltaic fuel cells are described as being of particular importance in remote areas without access to the main power grid. Reading against this, one might see that such research has more to do with expanding the energy market than with cleaner energy production. The confident inclusion of the PBMR in the sustainability section may seem odd to those readers who consider nuclear energy to have a question-mark over its contribution to sustainability of the Earth's ecosystems. However, it does not seem odd to a person whose subjectivity is constructed by a meaning of sustainability as sound development and good business sense. The projects to reduce the amount of electricity demand could be read as environmentally friendly, but is this because, as yet, ESKOM cannot keep up with demand? Given the hypothesis that ESKOM's definition of sustainability has more to do with business sustainability than with environmental sustainability, one might predict that it would not continue these projects should demand decline below its capacity to produce.

Explanation of the signs (social analysis)

> Different lexical selections can signal different discourses (colonial, liberal, labour). Most texts are hybrids, which draw on more than one discourse. ... Textual instantiations capture the clash of discourses and demonstrate ideological forces at work to produce a different hegemony (Janks, 1997: 335).

In this ESKOM document, the two main discourses are: the discourse of environmentalism (sustainability) and the discourse of economic rationalism (arguably very close to a discourse of neo-colonialism or "modernism", with its reliance on market-dominance expansion rhetoric and perhaps explaining why such images as the dark African continent seem natural). Although not very obvious in the *ESKOM Environment Report 2000*, there is also a hint of a patriotic/black-empowerment discourse (such as in the central position of Africa on the front cover).

The document should be seen within a context of strong pressure on companies to be "environmentally friendly" coupled, nevertheless, with a strong pressure to be successful within the dominant paradigm of economic development. The need for ESKOM to prove itself as an agent of development is especially strong because South Africa is a newly independent country with serious issues of poverty to address. There is a clash between these two ideologies and the discourses with which they are associated, but the contradiction in the use of both at the same time is hidden, for example, through the use of professionalism and etiquette.

Kress & Hodge (1988:3–5) explain how such an ideological complex as that which produced the ESKOM report "projects a set of contradictions which both legitimate and ameliorate the premises of domination". They explain how the different halves of the contradictions would cancel each other out. "We need, therefore, to invoke a second level of messages which regulates the functioning of ideological complexes, a level which is concerned with the production and reception of meanings." Kress & Hodge go on to explain that this second level of meaning must be "highly visible in politeness conventions, etiquette, industrial relations, legislation, and so on".

So, for example, in the PBMR contradiction where one discourse says nuclear energy is sustainable and the other says it is not, the contradiction is hidden in the following ways:

- in terms of rules about the production of meanings, it is hidden by the professional relegation of the reporting on the PBMR to the same importance as any other energy source. It would be "unprofessional" (or "overly emotional") to emphasise the PBMR issue in a serious report such as this one;
- in terms of rules about the reception of meanings, the contradiction is hidden by rules of politeness, as in, "it would be unfair and impolite to criticise ESKOM when it is clear that it has been trying so hard". Also, there are rules about criticising the "hand that feeds you": ESKOM has such a central role in producing power in the region that etiquette would have one politely allow its contradictions to go unnoticed. One risks being labelled unpatriotic or anti-development should one point out the contradictions in ESKOM's rhetoric.

The effect of these rules of production and reception is to legitimate and ameliorate the premise of domination upon which rests the position of ESKOM as a primary economic player in the region.

Parallel findings and deepening insights from the ESKOM Annual Report 2001

It should be noted that the *Environmental Report 2000* was written two years ago in a southern Africa that looked very different to the southern Africa of the present. September 11th, the economic and political strife in Zimbabwe and its effects on the economies of the region, had not yet happened. The New Economic Partnership for African Development (NEPAD) with its rhetoric of the African Renaissance was not yet in existence.

Perhaps as a result, the 2001 ESKOM annual reporting strategy looks very different to the 2000 *Environmental Report.* As suggested by the CDA of the 2000 report, which suspected the focus on environment was something of a publicity strategy, the change in international focus towards issues of immediate economic and national security and patriotism has meant that the environmental focus has been dropped. Instead of a separate environmental report, environmental reporting has now been included in the annual report. The 2001 report has no sections that deal directly with sustainability and only one heading dedicated to the environment, which simply gives the company's environmental policy. It would appear that in the clash amongst the different hegemonic discourses, the discourse around sustainability has been superseded by the discourse around the African Renaissance.

Parallel to the 2000 *Environmental Report*, nuclear energy in the 2001 *Annual Report* is given no more prominence than any other form of energy, and "sustainability" is used in ways that seem to imply sustainable business rather than a sustainable Earth. For example, the 2001 *Annual Report* talks of "sustainable energy supplies" (17) and "sustainable human resources" (25).

A characteristic of the African Renaissance discourse, as it has been interpreted in ESKOM's 2001 *Annual Report*, is the use of rhetoric borrowed from colonialist discourses. This seems to support the hypothesis that neo-colonialist rhetoric was present in the 2000 report. In the 2001 report, however, the rhetoric is stronger and more explicit. For example, there is a focus on large-scale projects and smaller projects to be replicated in all developing countries.

> A global effort is underway to design projects in South Africa ... using some of the core debates at the Summit as a foundation. This would be a global legacy that ensures that specific projects are implemented across Africa and are able to be replicated in all developing countries (ESKOM, 2001:22).

This is a characteristic pattern of neo-colonialist (some might call them "modernist") strategies: it is assumed that there is "one big answer" to everyone's problems. The possibility of locally initiated, highly differentiated projects at a smaller scale is overlooked. It is this assumption that leads to grand schemes in which small communities have few rights. These strategies have been severely questioned (e.g. Bauman, 2000). The assumption that one group of people can have "the answer" for another group of people underlies this model of development, and it was this same assumption of "knowing better" that lay beneath colonialist strategies of the past. Even the innocence of wanting to "help" and be of service, seen here in the *ESKOM Annual Report*, is commonly seen in the early African colonialist missionary rhetoric (Page & Page, 1991: 6).

A comment in the report that shows the will to be of service mixed in with the assumption of "knowing better" (bringing light into the darkness) is: "In a very real sense, ESKOM is a partner with other African utilities on the continent improving economies and advancing socio-economic development. We do this by both literally and metaphorically bringing light and energy to our continent" (ESKOM, 2001: 20).

Is this colonising rhetoric acceptable because it is being spoken by African people with "knowledge and power" about their "less-developed" countrymen/women and neighbours, rather than by foreigners about African people?

The report also draws on the unlimited growth model of rationalist (modernist) economics when the Chairman says: "It is now possible to visualise a different Africa ... an Africa whose scope for growth is limited only by its imagination ... (ESKOM, 2001:20)

This idea of unlimited growth is seen by some to be contrary to sustainability principles (Mies & Shiva, 1993; Sachs, 1999). The United Nations Environment Programme's (UNEP) *10 Years after Rio* assessment claims that while there are environmental improvements in industry, these are being overtaken by economic growth and increasing demand for goods and services (UNEP, 2002:19). Thus another question is: how viable is the unlimited growth model as the basis for sustainable development?

Conclusion

The ESKOM reports are instantiations of discourse. Their production and reception are governed by rules in which the writers of the texts are well-versed. Thus these texts are significant, in terms of this CDA, largely because of what they say about the discourses allowed by society (who is allowed to say what, when). In other words, the writers of these texts are governed by a set of rules about which they may be more or less conscious, but which nevertheless exert a strong control over them: they reflect the constraining and enabling rules of society.

Keeping this in mind, it is important that this analysis is not seen to be criticising the actual company of ESKOM or the writers of the annual reports. On the contrary (and this is why it is necessary to read "with" the text as well as "against" it): it is important to hear what the management is trying to say within the ESKOM reporting documents. Thus, it is true that ESKOM is one of the best examples in southern Africa of a company taking its environmental responsibilities seriously. It is also true that it is working hard to improve the quality of life for southern Africans in general: we could not do without ESKOM's contribution to society. Having personally worked in industry on environmental projects, I respect the work that ESKOM has put into its environmental initiatives. One of the reasons for choosing it to be the focus of this review is because is is so obviously above reproach, given the set of business rules by which it is constrained and on which its survival seemingly depends.

Bhaskar (1989:73) explained that such objects as ESKOM are not the starting points of interrelationships but the results of those interrelationships. He says, "on the realist and relational view advanced here, collective phenomena are seen primarily as the expressions of enduring relationships". Latour (1993; 1999) noted how he wanted to write without "denunciation" by focusing at the same time on both the network relationships and their apparent "quasi-objects" (such as ESKOM), rather than on the objects alone, or the networks alone.

Therefore, some questions that remain after this CDA might be: what characteristics of the interrelationships of society ensure that a company such as ESKOM, which has the intention to be more socially and environmentally responsible, nevertheless uses questionable rhetoric and a questionable business approach? ESKOM's rhetoric appears to mirror that of colonialist times and its business approach seems to draw on the same methods used by the people who created the current environmental crisis, despite evidence that those methods are seriously flawed. Another question might be: can sustainability, in its original sense, be achieved using the same approaches that led to the current socio-ecological problems?

This chapter has tried to show the complexity of environmental issues, and to break open for debate some of the hegemonic discourses currently being used within arenas focusing on sustainable development. The challenge is to create expectations of our businesses, such as ESKOM, that they move beyond their current globalising dominance rhetoric towards a power-sharing discourse, where the voices of the marginalised, and their call for greater environmental protection, are not silenced.

Endnote

1 KPMG is a global network of professional service firms providing financial advisory, assurance, tax and legal services (the letters KPMG stand for the names of the organisation's founding members).

CHAPTER EIGHT

Challenges for Environmental Journalism in Africa: A Case Story of NGO-based Journalism in Ecological Youth of Angola

Vladimir Russo, Angola

Successful environmental communication is not measured by public acceptance of the solutions formulated by decision-makers; it is achieved when the best solutions are chosen by a well-informed public. One of the most important roles of the media is to expand the audience for debate on a particular issue (Tolba, 1992).

Introduction

Ten years ago, at the United Nations Conference on Environment and Development (UNCED), some of the most important global environmental issues and a range of associated responses were discussed. A number of resolutions and conventions were approved and adopted. Key responses included *Agenda 21* (UNCED, 1992), and the NGO Principles for Equitable and Sustainable Societies (ICAE, 1993). Both responses recognise the important role that the media plays in creating public awareness and enabling the development of critical thinking and problem-solving skills. While these guideline documents presented some new ideas and approaches for environmental reporting, most of the recommendations for the media are aimed at targeting people with information to promote behaviour and attitude change. With increased realisation of the limitations of behaviour change orientations to education, awareness-raising and social change (Janse van Rensburg, 1995; O'Donoghue & Janse van Rensburg, 1995), it would seem to be important to review some of these assumptions in the arena of environmental journalism.

Agenda 21 is a global action plan for sustainable development, and puts strong emphasis on public participation in the decision-making process. It states that media should encourage public participation in discussions of environmental policies and that it should also ensure the dissemination of environmental information through mass media campaigns so that the information can be used by decision-makers at all levels (UNCED, 1992).

The *NGO Forum Treaty* (ICAE, 1993) emphasises the need to democratise the media as well as the media's responsibility to serve the interests of all sectors of society so as to promote wider public participation. It calls for the transformation of the media into one of the main education channels, focusing on disseminating information not only on an egalitarian basis, but also through exchanging values and experiences.

This chapter draws on a selection of issues associated with the role of the media in an African context. In particular, it will draw on a case study of *Ecological Youth of Angola* (*JEA*),[1] an Angolan NGO I established and worked with for more than a decade.

In *JEA* we drew on the principles in the *NGO Forum Treaty* to guide the development of our "communication for community" approach, as we tried to serve the interests of communities, and promote local community-based environmental action.

Before turning to a discussion of the *JEA* story, I briefly note some issues associated with environmental reporting in Africa, as I have observed these in clarifying a "communication for community" approach in the context of *JEA* over the past ten years. I had the privilege of representing *JEA* at the UNCED conference in Rio, soon after the NGO was established, and have followed the trends in environmental journalism since then, particularly as they have related to *JEA*'s work.

Some issues associated with environmental reporting in Africa

In general, the information disseminated by the media following the UNCED conference focused on a narrow understanding of "environment", often limited to ecological, biophysical and technical aspects. In many cases, environmental reporting was (and still is) limited to sensationalist headlines, followed by superficial interpretations of environmental issues. This problem is often compounded by the political interests that shape media coverage.

Other and related issues relevant to a discussion on environmental reporting include access to sources of information, subjectivity and bias in reporting, how stories are framed, the approaches taken to environmental reporting, and the priority afforded to environmental journalism.

Environmental journalism is often not seen as a priority. This influences the nature and amount of media coverage given to environmental issues. Even when environmental journalists are given space in which to report, they are often moved "upwards" to what are perceived as more prestigious reporting beats such as economics or politics (Keating, 1993:2). This further limits the continuity and quality of media coverage of environmental issues.

But are environmental stories indeed different from social, economic or political stories? Many journalists find it difficult to frame environmental stories within the available media frames. Environmental news does not easily fit a single "frame". For example, if a newspaper is reporting on global warming, it will fall under the scientific frame, a recent oil-spill under the actuality frame, a tree-planting campaign probably under the ecological frame, an environmental education meeting in the society frame. The question is: do we need to specify the frame used to report environmental issues or should environmental issues be recognised as being interdisciplinary and therefore open to consideration in the wider array of reporting frames? When environmental journalism focuses on either "green" or "social" issues without making the connection between ecological, socio-political and economic issues, the result is often shallow reporting. This may be due to a lack of insight into the complexity and diversity of environmental issues.

Often, the preferred frame for reporting on environmental issues is a narrow scientific one, and there is a tendency to aim to report "objectively" on facts and events associated with environmental issues. Other reporters go for what has been described as an event-orientation to journalism. For example, a daily newsletter established to cover the World Wilderness Congress held in South Africa in November 2001 reported mainly on the daily activities of the Congress and focused particularly on what "experts" said rather than on providing critical analysis of the contexts and issues being discussed. In our experience, this rather limited approach to environmental reporting fails to provide more contextualised and in-depth articles. By simply asking questions relating to the objectives and outcomes of events, journalists take for granted the existence of objective facts, a process which according to Hannigan (1999), leads to distortion and misinterpretation of environmental issues. Dunwoody and Griffin (in Hannigan, 1999:64) point out that event-orientated journalism "allows news sources to control the establishment of story frames ... and ... absolves ... journalists from attending to the bigger environmental picture".

In some countries, the media is still strongly influenced by authoritarian governments or dictatorship regimes, and reporters are forced to serve narrow interests as "old-paradigm journalism" filters, distorts and controls information, thus influencing environmental reporting possibilities (Thloloe, 1997:15). During the apartheid era, South Africa had a racially biased media, which was heavily controlled by the state. In Kenya, the government attempts to control news, particularly when it affects the economy or has the potential to contribute to political instability. Zimbabwe has also recently had problems with press freedom. State interference in freedom of the press had a profound influence on *JEA* while Angola was ruled by a single party. In the early 1990s, with the democratisation of Angola, old parties returned to the political arena and new ones were formed, putting pressure on the government to allow for more freedom of expression, which led to the opening of the private media (newspapers and radio stations). This freedom enabled *JEA* to establish its environmental programmes, voice the NGO's concerns and provide opportunities for people to participate in decision-making processes, with an equal voice. But it is not only in authoritarian regimes where press freedom may be an issue. The centralised control of media and media sources has motivated the establishment of independent media. The American Indymedia[2] use internet newswires and websites to disseminate information (Nogueira, 2001:26) in an attempt to decentralise the power from the mainstream media which is heavily dependent on government news sources (Edwards, 1998).

A related difficulty experienced by *JEA* and other environmental journalists in Africa is the sourcing of adequate information for their news, particularly where local people's perspectives on environmental issues are ignored. A series of television programmes developed by IUCN Mozambique[3] tried to address this problem by diversifying their sources of information and including opinions and perspectives from members of the community.

Approaches to environmental reporting: the challenge of going beyond one-way communications

There have been a number of critiques of behaviour-change approaches to environmental reporting. Taylor (1997:110), for example, noted that a "top-down" one-way form of communication may not only alienate the public and prevent meaningful action from being taken, but could also fail to draw on the rich diversity of available knowledge of issues and risks in different contexts. I want to argue for an approach to environmental journalism that can be called "communication for community" as it attempts to involve communities in the reporting process and, at the same time, involve the media as members of the community in local responses to environmental issues and risks (Russo, 2001).

I shall illuminate this approach by sharing some of the approaches to environmental journalism that the members of *JEA* have explored in the last ten years. I would argue that a key challenge for both NGO-based and mainstream environmental journalists would be to conceptualise approaches and methods that engage a community focus.

Some of the issues that we have tried to concentrate on at *JEA* include:

- balanced reporting (reporting objectively, while recognising subjective factors);
- reporting using fewer technical details (clarifying the issue without oversimplifying it) with comments and participation from the community;
- open-ended suggestions that encourage people to reflect, draw their own conclusions and make their own judgements;
- sharing good stories and successes (a focus on action-taking) by supporting different community initiatives and getting close to people affected by environmental problems.

Through a description of our case study and some of the issues we have tried to confront over the past ten years, I wish to explore dimensions of our communication for community approach to environmental journalism. Lawrence (cited by Berger, 1996:29) notes that journalists should focus on stories of real importance for their communities, and that journalism should tell people what is really happening by focusing on meaningful and relevant stories for the community and "telling them thoughtfully". Journalists should not be afraid to celebrate and report good practices and achievements. It is in this spirit that I "tell our story thoughtfully", and hope that the community of environmental journalists will find this a relevant story to stimulate further debate on approaches to environmental journalism, particularly in the light of the obvious need for the media to support informed public participation and community-based responses to environmental issues and risks in Africa.

The case of Ecological Youth of Angola

For the past ten years, *Ecological Youth of Angola* (*JEA*) has focused on three main areas: environmental education activities, capacity-building and media programmes (print, radio, television, and latterly the internet). This case reports mainly on the media-related work of *JEA*.

Green Phone aired for the first time in 1995. It was the first environmental radio programme in Angola. The name was based on the initial idea of having a phone-in live programme enabling people to ask about, and report on, environmental issues. As a result of its success, importance and sincerity (by providing appropriate information, supporting action-taking initiatives and enabling people to express freely their specific local problems and concerns) Green Phone was broadcast for six years and won awards for its contribution towards the improvement of people's quality of life.

Other radio programmes created by *JEA* include Green Channel (in 1998) and Blue Planet (in 1999). In an attempt to enable broader local responses, Green Horizon (first airing in 1999) and Communicating with the Environment (in 2000) were created in two provinces not linked to the capital Luanda. In a context of war and socio-economic instability, these radio programmes tried to establish a culture of "communication for community", in an attempt to promote or provide practical support to action-taking initiatives that would enable people to respond to issues, and contribute to improvements in quality of life at a local level.

JEA has a strong action-competence approach to journalistic activities. All our radio programmes encourage people to present their opinions, concerns and problems. Possible solutions are always offered, and where necessary referrals, to other people or institutions, are provided. Reporting on successes has been one of the key principles of *JEA*'s media programmes, in order to share lessons and learn from the past. A number of environmental issues have been addressed from both social and economic perspectives and we have also taken an active role in mediating environmental conflicts.

Since these programmes were implemented by an environmental NGO, there was a strong advocacy role evident in them. Journalists producing the programme were members of *JEA* and their advocacy role was to lobby for and support local initiatives. The commitment of *JEA*'s members to be responsive to people's concerns (within a framework of a broader view of the environment) led to lively debates with local authorities and government officials, often with positive results for the communities affected by problems. The involvement of civil society was a key feature of all the programmes, not just because the public was given voice (or air time) but also because of the establishment of discussions on issues where journalists and community members were able to debate and consider possible solutions to local problems. This approach tended to involve the journalists in "hands-on" approaches to engaging with local communities, both in the reporting of issues, and in the action-taking.

In both radio programmes and newspaper articles, there seemed to be very few problems with editors understanding the environmental issues being reported. This was probably a result of the small-scale operations of *JEA*, and the fact that our members are all committed environmental activists. There is also an educational component to our activities, which allows us to become involved in a number of training activities on environmental issues. *JEA* journalists are therefore actively involved in broadening their own understanding of environmental issues and risks, with a view to enhancing their environmental reporting skills as well as their capacity to support local environmental action in community contexts.

Recently, the United Nations Environment Programme (UNEP) recognised *JEA*'s work and awarded the NGO with the prestigious Global 500 Award during the celebrations of World Environment Day in Shenzhen, China, in June 2002.

Establishing the orientation to, and focus of, *JEA*'s journalistic work has taken ten years. In the discussion below, I shall highlight some of the key issues we have confronted and considered in establishing our "communication for community" approach. In doing this, I draw on the perspectives of different authors commenting on environmental journalism practices, to locate our work in a broader social context, and in the broader arena of environmental journalism practice.

Framing our practice

One of the challenges we faced was the need to clarify the frame used in environmental reporting. In order to be able to recognise diverse orientations, cultures and understandings of the environment, we adopted an interdisciplinary approach, with a view to establishing broad-ranging debate in our community interactions.

We found that a key issue linked to establishing our frame was the need to establish a broader understanding of the environment. Drawing on the work of O'Donoghue and Janse van Rensburg (1995), we developed a perspective on the environment as comprising interactions between different dimensions: social, biophysical, economic and political. We also recognised the complexity of the interactions amongst these dimensions, and we recognised that journalists often seem to have difficulty setting up a particular frame. From our experience, we noted that a better understanding of the concept of "environment" can help journalists to frame stories and news differently. We found this to be an important dimension of our training programmes, and as journalists in *JEA*, we regularly sought information and considered debates about the inter-relationships between different aspects of environmental issues and risks.

We also found that this broad view of the environment helped editors demystify the idea that environmental issues were only biophysical in nature (soil erosion, desertification, acid rain, trees and animals). This also provided a conceptual framework for challenging some of the more scientific perspectives on environmental issues, and enabled us to broaden our environmental coverage. To share this broader perspective of the environment with other journalists, *JEA* members started a newsletter in 1996, produced a number of articles for newspapers and invited journalists to attend courses on environmental issues. As a result a number of journalists started to broaden their scope and include political, economic and social aspects in environmental news.

Commitment

A further challenge for *JEA* in clarifying our orientation to environmental journalism has been the need to confront assumptions concerning neutrality or objectivity in environmental reporting. A principle in the *NGO Forum Treaty* suggests that the media should be "committed to serve the interests of all sectors of society" (ICAE, 1993). The challenging question we considered was just what this might mean in an environment where political freedom had been curtailed and where media interests were shaped by government interests. How committed should the media and journalists be, and in what ways should they show their commitment? What kind of commitment is needed in a democratising society? What does "all sectors of society" mean in the Angolan context? We considered it important to recognise our social responsibilities and our commitment to democratic values and social upliftment in Angola. We articulated this commitment by realising that we could not show our commitment by providing news to raise awareness only, or by simply reporting on government interests and priorities. We had a commitment to share information in ways that would promote change and address people's environmental concerns. In exploring what the meaning of commitment in environmental journalism might be, we also drew on insights from an environmental journalist from Brazil, Liana John (2000:4), who noted that journalists should also assume a commitment in relation to the objectivity, clarity and views that are reflected in their reporting. Thus, while we made a clear choice about our commitment, we were also careful to consider issues of objectivity, as I'll discuss further below.

Targeting recipients, or learning together?

The *NGO Forum Treaty* (ICAE, 1993) recognises that we are all learners and educators. We found this to be an important guiding principle in our work. In clarifying our "communication for community" approach, we realised the need to be involved in sharing ideas and solving environmental problems with others. We did not have all the answers, and so we tried to find ways of involving community members and experts in our environmental journalism. This approach is being used with increasing frequency in different community-radio initiatives in Madagascar, Mozambique and South Africa.

In challenging a belief that environmental journalism (like other mainstream journalism), is about "getting the message across", we have found that it is important to challenge the idea of a "target group". Paolo Freire (in Linney, 1995:6) noted that target-based approaches are often based on the "myth of people's ignorance". We have noticed that target-based approaches often lead to alienation and anger in the target group. These approaches seldom promote dialogue and participation and are therefore limited in enabling public participation in decision-making processes. We have tried, instead, to consider the audience of our work as partners rather than targets. Based on a developing understanding of the nature of two-way communication and the implications of "communication for community" *JEA* considered civil society a partner in our attempts to respond to environmental issues and risks in Angola. This led to experimenting with different ways of engaging people in our journalistic activities. Members of civil society were often invited to our radio programmes to share their views on a particular problem or to help us to decide which topics to report on in the next programme, or to discuss action projects and possible solutions to problems.

Palmer (1998:275-276) noted that education and communication processes often seek to change perceptions of "their target audiences" by providing better and more reliable information, which seems to be insufficient. She notes that "... knowledge of facts alone ..." does not bring behaviour change and it is important to link information to personal benefit (Palmer, 1998). Adopting social marketing approaches in environmental journalism promotes "communication as persuasion", and involves getting to know people's needs so that messages can be sold to them. These approaches often involve using popular culture (e.g. music idols) to help sell the message. In considering this approach, *JEA* journalists discussed some of the assumptions of social marketing approaches, which include the assumptions that the communicator (or expert) is able to determine what message is best for the other (target audience), and that the communicator has only to learn what the other appears to need, for the messaging to be successful. While slightly different from earlier targeted-message approaches, social marketing approaches still tend to treat civil society as recipients of their messages, with the messages being made more appealing. In *JEA* we recognised that some of the strategies used in social marketing approaches might be useful to enhance interest in media messages, but we tried to maintain our commitment to engaging civil society in interactive dialogue, and our commitment to supporting local action initiatives and the reporting of success stories.

Objectivity

A further challenge for environmental journalists is to consider what they report, and how the reporting is done. This, like the way journalists view and involve their audience, is often associated with ethical issues. Keating (1993:92) considers objectivity a crucial aspect of environmental reporting, arguing that journalists should try to be "disinterested observers". However, he notes that journalists should also recognise that they "... have a duty to fairly and accurately explain risks and alternatives to people". For Morris (1996) objectivity is the journalist's commitment to balance, fairness and independence of views in reporting.

It is often assumed that, in a democratic country, reporters will be objective. In reality, they often end up serving the interests of the corporations that control the media, political systems and the socio-economic framework they want to promote. These "biases" often compromise objectivity. For example, companies may withdraw advertising if papers run environmental stories which challenge their actions, or multinationals might censor certain information if they are funding environmental projects. In Angola, an oil company supporting the production of a music CD, "requested" of an Angolan singer that he rephrase the lyrics of one of his songs because it mentioned the effects of an oil spill (caused by the same company). For Fishman (1977:52-4) this is manipulated journalism within the political game designed to serve particular interests. *JEA* tried to overcome these challenges by acting as a facilitator, and used strategies to stimulate debate rather than push particular perspectives. In this way, we were able to encourage critical thinking skills.

And even when journalists are free of any constraints on what they report on, are their personal views ever objective? Should they be? Can they be? Without having an opinion on the facts it is difficult to report meaningfully; however it is important to be rigorous and honest. In *JEA* we had numerous debates on how to increase our objectivity (e.g. by knowing more about the issues, considering the issues from different perspectives, stimulating debate) while recognising our own subjectivity and value orientations.

To Killingsworth and Palmer (in Hannigan, 1999:67), reporting using objectivity and balance means that journalists "... often attempt to distance themselves and their readers from the environmental struggle ... taking refuge instead in the objectivism of science". Thus, journalists seem to express themselves "... as a neutral and ironic voice ... that ... rarely express[es] the content of environmental stories in overtly political terms, opting instead for news frames which emphasise conservation, civic responsibility and consumerism" (Hannigan, 1999:67). In our radio programmes, for example, we have tried to report objectively while drawing on our knowledge and experience of the context (thus being subjective), and the knowledge and experience of local people. Adopting an advocacy role, while trying to be as objective as possible, was often quite a challenge for our work, particularly as environmental issues are often complex and contested. The discussion surrounding objectivity and subjectivity is likely to continue to be a difficult area in journalistic work.

Enabling participation and action

In both media programmes and educational activities, *JEA* colleagues were committed to environmental learning and action for change through community involvement in interactive debate, critical questioning, problem-solving and action-taking. To support action-taking, *JEA* has promoted a spirit of sharing ideas through cooperation rather than simply reporting news. We found that this approach could stimulate open-ended debate on environmental problems and enable action and collaboration between the community and the media (Russo,

2001). For example, *JEA* has supported a number of community-based environmental projects such as tree-planting campaigns, re-utilisation of paper and metal, and the establishment of nurseries.

The action focus of *JEA*'s media work was enhanced by the use of critical communication techniques and socially critical approaches. Fien (1993) has argued that a socially critical approach to information may enable action-taking. Where possible, we used radio and television techniques that encouraged partner groups to question us critically, and we tried to stimulate critical analysis of news events. For example, debates on how to explore alternatives in addressing environmental problems, critically justifying our views and attitudes towards sustainable development, and promoting partnerships looking for change were evident in our media programmes. We found that these approaches helped the audience to draw on "what they already knew", and to ask questions that they may not otherwise have thought about.

Mobilising communities' (prior) knowledge

In our work we noted that sources of environmental information in Angola were confined to small groups such as specialised public agencies and environmental NGOs, and to academics or scientists. We found, for example, that promotion of particular forms of knowledge construction (which are often narrow and limited) was visible where the sources of information were powerful and controlled by the government or large corporations who continue to repeat the same message (Russo, 2001). For example, in one of *JEA*'s radio programmes broadcast by state radio, environmental news could not reflect any message contrary to the government's messages. The strategies noted above, to report on success stories, and to draw on local knowledge, were ways in which *JEA* tried to overcome these problems.

We also tried to overcome these limitations by mobilising communities' prior knowledge in investigating environmental issues.

However, mobilising community knowledge does not rule out challenging communities' understandings (which may be inaccurate or fail to support change towards more sustainable environmental management or lifestyles). Therefore, we believe that it is important to bring local perspectives into conversation with experts (through interactive debates or phone-in programmes). We have found that these kinds of interactive debates often enable people to see things differently, and that these processes are educative in the sense that they provide people with new understandings and insights which can be applied in local contexts (Russo, 2001). We have found that we are also challenged to review our own perspectives on issues, through these approaches to our journalistic work.

Reporting successes

Reporting on successful case studies and environmental stories helps to motivate people and encourage action. For example, when we reported on a successful project combating desertification in southern Angola, a number of people added their support to the project and various requests for further information were received by *JEA*. Reporting on failures and overwhelmingly negative issues can lead to apathy.

Rather than criticising mistakes from the past, *JEA* has concentrated on celebrating successful initiatives on sustainable development and enabling the sharing of ideas between countries. Organisations such as the Wildlife and Environment Society of South Africa and Environmental Education Association of Southern Africa (EEASA) have also been trying to report successful environmental projects, thus promoting networking and sharing of ideas regarding environmental action.

Prioritising communication: communication for community, not persuasion (or indoctrination)

In attempting to clarify our "communication for community" approach to environmental journalism over the past ten years, we have moved away from more traditional "top-down" forms of environmental journalism, and from linear models of communication.

We have tried to move away from processes that simply try to "get the message across". Relationships between journalists and communities are not linear. Target-based approaches to communication have been critiqued for limited understandings of processes of social change (Taylor, 1997:110).

Thus, over the past ten years, *JEA* has developed a "community-based approach to journalism", which aims to work with people in local contexts, and to create opportunities for local dialogue and debate. We have called this a "communication for community" approach, noting that the word "communication" (like "community") has its root in the Latin *commune,* which means to share, to be in close touch with other people.

Conclusion

Adopting an advocacy role in journalism is challenging. The reward for journalists is being able to play a proactive role as citizens who have the potential to improve aspects of people's quality of life. However, our task seems to be a complex one as, in advocating for the environment, we might be seen as extremists, which might be confused with radical environmentalism (such as some of the Greenpeace activities or anti-globalisation movement protests). What is encouraging is that a growing number of institutions in Africa, such as Media Institute for Southern Africa (MISA) in Zambia, and *Eco News Africa* in Kenya, are following a similar path and using structures and publications to advocate for people and the environment.

We have learned to engage the arguments around objectivity and subjectivity in journalism by being conscious of the issues of accuracy, rigour and bias, while showing commitment to the principles of democracy and advocating for environmental and social change. We have found this to be an important area of consideration in relation to professional ethics. Are we ready to move from our comfort zone to a more challenging and proactive role as environmental journalist-activists?

The *New Internationalist* (2001:18-19) notes that fewer and fewer media giants are controlling more and more of what we read, see and hear. The monopolisation and centralisation of media power within the same circuits will no doubt have a great impact on environmental reporting in future. We see the need to continue exploring partnerships with other media groups, NGOs, governments and local communities as an important area for growth. A combination of environmental activism, communication skills and partnerships with other media organisations and communities has strengthened our work.

Based on our experience in Angola, providing a meaningful framework for environmental reporting, which is based on a concept of "communication for community", broader perspectives of environment, multiple and diverse sources of information, mobilisation of communities' knowledge, advocacy for the environment, and engagement in action-taking, is not an easy task. It is important to consider that besides the general challenges of journalism (deadlines, lack of support, dominance of event-orientated approaches), environmental journalists often work under additional constraints and usually in an unsupported fashion.

However, despite the limitations, environmental reporters work in an exciting arena. The biggest-ever global summit is taking place in Africa in 2002. Over 5 000 journalists have already been registered for participation in the WSSD. How journalists will take forward the issues raised during this summit, is a key challenge for the profession.

Endnotes

1 I use the word "our" when referring to the work that journalists at JEA – including myself – have undertaken together over the past ten years.

2 Indymedia – The Independent Media Center – is a global initiative that was created because of the "... frustration with the way mainstream media sensationalises or ignores the almost daily anti-globalisation protests occurring around the world". (www.indymedia.org)

3 The World Conservation Union brings together states, government agencies and a diverse range of non-governmental organisations in a unique world partnership. It has members spread across some 140 countries (including Mozambique).

CHAPTER NINE

Curriculum Patterning in Environmental Education: A Review of Developments in Formal Education in South Africa

Heila Lotz-Sisitka, South Africa

> For education activists in Africa and elsewhere, there is a growing challenge and an urgent need for reconceptualising education as a whole, and articulating alternative scenarios and frames of reference ... the foundations of education systems [in Africa] require a comprehensive overhaul if [they] are to begin to respond to the contexts within which [they] are located (Odora, 1994:96).

Introduction

This chapter reviews trends and challenges in environmental education curriculum development processes in South Africa. It draws on empirical data from 1992 to 2002, highlighting the interaction between environmental education and national curriculum development initiatives. Trends, challenges and tensions are reviewed in the light of a perspective which views environmental education as a process of social transformation against a backdrop of international developments following the 1992 Earth Summit and, in particular, post-apartheid transformation. This latter process provided unique opportunities for environmental educators to explore the social transformation role ascribed to environmental education processes, within the institutional frameworks of school curricula and formal curriculum development processes. Reviewing a ten-year period of extensive curriculum work can be approached from many directions. Against the above, I focus this review on the role that environmental education curriculum development processes may play in enabling greater reflexivity in response to environmental issues and risks. My underlying question is whether we are establishing curriculum processes that affect the agenda for social change.

The chapter draws on a review of empirical data and research associated with policy and curriculum development processes linked to the transformation of the formal education system in South Africa. I draw most strongly on the following initiatives:

- the Environmental Education Policy Initiative (EEPI) – which between 1992 and 1995 introduced a participatory policy-making process to environmental education curriculum work in South Africa just prior to, and after, the first democratic elections (Clacherty, 1993a; 1993b; 1995; O'Donoghue, 1993; Lotz-Sisitka & Ashwell, undated);

- the Environmental Education Curriculum Initiative (EECI) – a state-civil society partnership project (1996–2000) enabling staff from the Department of Environmental Affairs and Tourism, provincial government education departments and environmental education practitioners around the country to work together to debate and define environmental education in the emerging new school curriculum, known as Curriculum 2005 or C2005 (Janse van Rensburg & Lotz, 1997; Lotz-Sisitka & Ashwell, n.y.p.);
- the Learning for Sustainability Pilot Project – a donor-funded pilot project (1997–2000) which focused on the professional development of teachers to enable them to enhance their skills for learning programme development in a context of rapid curriculum change – in two provinces (Bjerg-Hansen & Sguazzin, 2000; Du Toit & Sguazzin, 2000; Janse van Rensburg & Lotz-Sisitka, 2000; Lotz-Sisitka & Janse van Rensburg, 2000);
- the National Environmental Education Project for General Education and Training (NEEP-GET) – a large-scale donor-funded initiative (2000–2002) aimed at providing professional development to curriculum advisors and teachers to enable the integration of environmental learning in schools. This project operates in all nine provinces (NEEP-GET Project Document, 2002; Lotz-Sisitka & Raven, 2001).

The above initiatives represent the major national curriculum intervention in environmental education curriculum development work between 1992–2002. Post-apartheid political transformation gave rise to a radical re-orientation of the role of education in South African society. The chapter reflects the political role played by environmental educators in conceptualising and re-orienting environmental education in the context of socio-political transformation in the country.

International perspectives influencing environmental education curriculum development work

Since the 1992 Earth Summit in Brazil, many African countries have signed a number of multi-lateral environmental agreements, including Agenda 21 and numerous environmental conventions, which influenced local policies and practice, including education. In South Africa, the country's new Constitution enshrined the right to a healthy environment as part of the Bill of Rights, and emphasised the concept of sustainable development in national policy discourse.

Despite increased international attention and governmental interventions, societies around the globe are aware that local and global responses to most environmental issues are not sufficient at present. A recent *Living Planet Report* (WWF, 2000) concludes that the state of the Earth's natural ecosystems has declined by about 33 per cent over the past 30 years, and that the ecological pressure of humanity on the earth has increased by about 50 per cent, which is said to be exceeding the biosphere's regeneration rate. Despite unprecedented economic growth on a global level, poverty levels are increasing and patterns of environmental risk are becoming more complex (Beck, 1999). At the start of the new millennium, more than 125 million of the world's children are still deprived of any schooling while another 150 million drop out of school without learning to read or write (Brazier, 1999; see also Odora, 1994 for a critical comment on education in Africa).

There is also growing awareness of the limitations of knowledge systems, and a loss of faith in the certainties of science (Beck, 1992 and 1999). We are increasingly aware of our limited capacity to predict or determine exactly what our impact on the environment might be (as recent debates on genetic modification of seed illustrate). Nor are we able to establish accurately what the earth's capacity for sustaining life indefinitely might be (witness discussions on the precautionary principle, e.g. O'Riordan & Cameron, 1994). These realisations have had profound implications for the epistemological and economic certainty established in the 20th century under the guise of modernism, and we start this millennium with enormous challenges relating to the need to re-orient and reconceptualise the very modernist foundations on which development – as well as education systems and curricula – has been based. The basic premises of dominant models of development based on unlimited economic growth and the globalisation of culture and trade are also being questioned (Burbules & Torres, 2000).

These questions include critiques of linear, narrow and mainly economic premises on which the notion of progress is based (see for example Shanin, Sachs, De Senarclens & others, in Rahnema, 1997). Increasingly we are realising that historical change is not structurally determined or predetermined by any laws or models of social change, but shaped by complex and unpredictable dynamics of social struggle in new time-space configurations (Cross, 1999; Popkewitz, 1999). It would appear therefore that we need to think of change as taking place within horizons of possibilities different to those established by modernist framings of social life.

Popkewitz (1999:17) notes that contemporary (globalising) society is characterised by

> important struggles in knowledge production … which point to significant epistemological ruptures in the doctrines of "reason" which have dominated disciplinary knowledge and social practice since the late 19th century (including education and curriculum; see also O'Donoghue, 2002; O'Donoghue & Neluvhalani, this volume).

These developments in social theory may indicate a radical change in the way humans are beginning to view their societies. Bauman (2000:23) quotes Giddens who reminds us that we are all engaged in "life politics"; we are "reflexive beings" who look closely at every move we make. Bauman, however, points to the limits of this reflexivity when he notes that

> … reflexion does not reach far enough to embrace the complex mechanisms which connect our moves with their results and decide their outcomes, let alone the

> conditions which hold such mechanisms in full swing ... our critique is unable to affect the agenda set for our "life political" choices.

Thus, in reviewing tensions in curriculum development work over the past ten years, I am guided by the following questions: what role could environmental education curriculum development processes play in enabling greater reflexivity in response to environmental issues and risks? Are we able to establish curriculum processes that affect the agenda set for life political choices? Can environmental education processes be processes of social transformation?

Developments in the field of environmental education

In the last ten years, the scope of environmental education has expanded with increasing emphasis on the role of education in responding to wide-ranging, complex environmental issues and risks. Chapter 36 of *Agenda 21*, and the *Treaty on Environmental Education for Sustainable Societies* (adopted at a plenary Earth Summit meeting by the International Forum of NGOs and Social Movements on 12 June 1992) both recognise the central role of education in shaping value orientations and social action. Both documents describe environmental education as socially transformative, continuous learning processes based on respect for all life. The NGO Forum principles have become influential in post-apartheid South African environmental education circles, as they show a concern for social justice, equity, democracy and social transformation. These principles support a dynamic and open climate of cultural interchange and the need to consider social and historical contexts of environmental issues – processes which create openings in environmental education for the recognition of diverse cultural values and indigenous knowledge, and a deeper understanding of the root causes of environmental issues. The NGO Forum principles also create a more democratic framework for educational processes (including curriculum development work), recognise the socially constructed nature of knowledge, and foreground the need to promote, recognise and respond to cultural, linguistic and ecological diversity (ICAE, 1993).

These principles, and other models influencing the re-orientation of education and training systems worldwide towards sustainability, have in turn been widely influenced by socially critical orientations to curriculum (articulated by Freire in the context of educational empowerment and transformation in the 1970s, and by Fien, 1993; Greenall Gough & Robottom, 1993 and others in the context of environmental education in the early 1990s). Socially critical orientations to curriculum became popular in South Africa in the 1990s, particularly in environmental education (see for example Lotz, 1996; Klein 1997; Wagiet, 1998; Du Toit & Sguazzin, 2000) and were viewed, along with other versions of critical theory (inspired by neo-Marxist philosophies of consciousness), as possible routes to empowerment and emancipation. The influence of neo-Marxist versions of critical theory in environmental education has, however, been critiqued for ethical (Jickling, 1999) and epistemological reasons (see for example Ellsworth, 1989; Janse van Rensburg, 1995; Popkewitz, 1999), particularly as critical theories were adopted as "crusading rhetoric" in attempts to empower the (less powerful) "Other".

Janse van Rensburg (1995), for example, in viewing environmental education as a process of social transformation, recommended not only critical perspectives, but a reflexive perspective or orientation to change, which emphasises process rather than product, which is not concerned with a linear model of change, and which does not rely on doctrines, tools or methods to bring about change. She sees environmental education as a "responsive process of change" involving the development of the capacity to "collaboratively develop capabilities (tools, resources, action competencies) to deal with and encourage change in local contexts" (Janse van Rensburg, 1995:168).

Despite global efforts at establishing an agenda of social change, widespread concern about the state of the environment, and the placing of environmental education on the political agenda for over two decades, the field continues to operate mainly on the margins of the formal education arena (Gough, 1997:142). In the mid-1990s Robottom (1996) argued that it might well be better for environmental education to remain permanently peripheral, particularly as environmental education processes are contextually linked to diverse issues, and that responding to these issues requires critical and reflexive skills which are not easy to institutionalise in mainstream curricula. His concern is for the way in which institutionalisation may lead to centrally indicated, decontextualised content for environmental learning, and assessment of technical knowledge and skills, a perspective picked up by Le Grange and Reddy in the South African context in early comments on integrating environmental learning in mainstream curricula (1997).

In many African countries, environmental education is seen as peripheral to mainstream curricula, particularly in those education systems still influenced by colonial education frameworks (Mokuku, 2000; Mokuku, this volume). Odora (1994) has argued that colonial (Western) education has subjugated, marginalised and fundamentally changed environmental knowledges in Africa.

In South Africa, the post-apartheid transformation of education provided unique opportunities for environmental educators to explore the social change role ascribed to environmental education processes, within the institutional frameworks of formal curricula and policy processes, a journey that has been characterised, as reported here, by many tensions and challenges.

Curricula as processes of patterning

Stenhouse (1975:1) notes that the *Shorter Oxford Dictionary* has used the term curriculum since the 17th century, which, he notes, marks the beginning of systematic and self-conscious attempts to regularise courses of study. He notes that there are many different views of a curriculum, notably "curriculum as an intention, plan or prescription, an idea of what one would like to happen in schools", and as "the existing state of affairs in schools, what does, in fact, happen" (Stenhouse, 1975:2).

In recent years, curriculum scholars such as Grundy (1987), Cornbleth (1991) and Doll (1993) have formulated critiques of instrumentalist discourses within curricula such as those proposed by Tyler (1949). Drawing on Habermas's knowledge-constitutive interests, Grundy (1987:6) presents the curriculum as a cultural construction, concerned with the experiences people have as a consequence of the curriculum, rather than with its constitutive aspects. She argues for a view of curriculum which centralises the making of deliberative judgements and the meaning-making function of the educational enterprise, rather than the production of predetermined outcomes (Grundy, 1987:99). Cornbleth (1991), also drawing inspiration from critical theory, argues for a process orientation, and sees the curriculum as a contextualised social process. These perspectives tend to define curricula not through reference to universal schemes and principles, but according to particular judgements of situation and circumstance. Goodson (1994) contributed to this critical discourse by noting that not only should one consider the culture and context of the curriculum, but one should also consider the kinds of political and epistemological parameters and interests at play in its construction. Doll (1993), in providing a broad-ranging overview of the epistemological parameters of contemporary society, envisions a postmodern curriculum. It will allow creative organisation and reorganisation of the educational experience in certain environments. It will also maintain a healthy tension between the need to find closure and the need to enable a system which allows for open-ended, critical and reflexive dialogue and deliberations between teachers and learners.

In his consideration of political, social and epistemological influences on education, Bauman (2001) describes the social function of institutions of learning as being crucial sites for a patterning of the values which are instrumental for social integration. He sees the role of the educator (in modernist forms of education) as translating these values into social skills, and therefore ascribes a phenomenon of social patterning to education, in which individual members of society are shaped into social beings "...fit to perform, and willing to abide by, the socially assigned roles" (Bauman, 2001:129). Educational institutions and their role in society, he notes, are a modern creation, concerned with the marriage of knowledge and power. In this vein, Cross (1999:298) notes how the "logic of difference" characterising dominant discourse in South Africa in colonial and apartheid regimes translates into a politics of separation, differential treatment and segregation in education. Apartheid curricula, based on an oppressive value system which stressed racism, sexism, tribalism, individualism and elitism, have led to a patterning of identity based on racist and exclusivist images of identity. In redressing this pattern, South African education will require a "...flexible policy that will enable learners to master the skills of intellectual production, and the use of these skills in active critical self-consciousness" (Cross, 1999:298), thus indicating the need to give critical attention to patterns of identity construction and life-political choices. In the contexts of uncertain patterns of environmental risk and the space created by post-apartheid transformation in education, curriculum patterning in South Africa clearly requires considerable thought from socio-political, socio-economic and socio-ecological perspectives.

As indicated above, patterning in education takes different forms, often linked to dominant discourses, socio-political decision-making, and the social functions ascribed to education. In the past ten years of engaging with environmental education curriculum development processes in South Africa, I have observed different kinds of patterning taking place, all of which have significance in the context of post-apartheid transformed education, and in the context of the social change role ascribed to environmental education. I have clustered some of these patterning processes inherent in curriculum development work as follows:

- participation;
- the visible curriculum (curriculum policy documents and assessment frameworks);
- processes of learning (expressed in the defining and shaping of learning theories and methods);
- reflexive processes in environmental education curriculum work (emerging attempts to deepen our understanding of curriculum issues).

Patterns of participation

Changing political economies of curriculum development

One of the earliest areas of tension was the hotly contested arena of participation in the curriculum development process itself. In South Africa, education has been at the centre of political struggle for many years. With the unbanning of liberation movements and the release of political prisoners in the early 1990s, the political environment changed dramatically. Policy-making came to be reinterpreted through the language of participation, transparency and accountability. A key player in the mass democratic movement at this time was the National Education Coordinating Committee (NECC), who initiated the National Education Policy Initiative (NEPI). The NEPI proposed a number of policy options to the new government, including curriculum policy options. A key recommendation was the establishment of curriculum structures which allowed for "greater openness and participation" (NEPI, 1993:107), noting that the "future curriculum would be developed and changed within the broader political context" (NEPI, 1993:115).

Following contestation which saw the 1987 *White Paper on Environmental Education* being rejected as illegitimate because of a lack of participation, and a realisation that formal education curriculum policy had to be made within the formal education arena, the Environmental Education Policy Initiative (EEPI) (a participatory state-civil society policy-making alliance) was established in 1992. This process resulted from a Policies and Procedures survey coordinated by the Environmental

Education Association of Southern Africa (EEASA) in 1992, in response to the desire to play a more proactive lobbying role in educational transformation (EEASA, 1992). One of the earlier tensions associated with this initiative was the style of policy development, where some participants in the EEPI were more comfortable with a tightly structured and efficiently managed hierarchical approach to policy development, while others favoured a less rigid, more participatory and open-ended one (Clacherty, 1993b). The EEPI working group opted for political alignment and a more democratic approach, arguing that practitioners had to be involved in the process because "policy is nothing unless it includes practice" (Clacherty, 1993b:4). A key outcome of the deliberations of the EEPI was a resolution at the NECC conference in 1993 which stated that:

> This Conference ... therefore resolves that ... the curriculum will develop the understanding, values and skills necessary for sustainable development and an environment that ensures healthy living (Clacherty, 1993b: Appendix 17).

This clause was reproduced in the ANC's policy framework for education and training in 1994 (ANC, 1994), leading to the principle in the 1995 *White Paper on Education and Training* (Department of Education [DoE] and Department of Labour [DoL], 1995), which set the scene for ongoing environmental education curriculum development work:

> ...environmental education, involving an interdisciplinary, integrated and active approach to learning, must be a vital element of all levels and programmes of the education and training system, in order to create environmentally literate and active citizens and ensure that all South Africans, present and future, enjoy a decent quality of life through the sustainable use of resources (DoE and DoL, 1995:18).

Thus, through the participatory policy-making process, environmental educators established themselves as stakeholders in the curriculum development process in South Africa, participating in the subsequent development of C2005 (DoE, 1997a), the streamlining and strengthening of this curriculum, and the development of a revised national curriculum statement (DoE, 2002).

Almost ten years later, the National Qualifications Framework (NQF) study team (DoE and DoL, 2002:70) notes that "stakeholder participation has been etched into the democratic psyche of a large segment of the SA society" and that while this is often both expensive and a major reason for slowness, it is also an important dimension of the transformation process.

However, this process is not without considerable challenges. Significant among them, from a curriculum perspective, is that participation of/as "stakeholders" does not necessarily lead to participation as inter-subjective meaning-making in context.

There is considerable evidence[1] to indicate that in spite of wide-ranging participation by teachers and other stakeholders in curriculum development and implementation in schools and classrooms, this has not made the process any easier. O'Donoghue's (1999) review of participatory trends in environmental education processes in South Africa, noted the often paradoxical ethical inversions and moral imperatives which have become characteristic of participatory approaches to curriculum development work.

Experience with C2005 points to the way in which complex curriculum frameworks can obscure, decontextualise and disable meaningful participation in curriculum work in schools (Lotz & Olivier, 1998; Taylor & Vinjevold, 1999; Committee for the Review of C2005, 2000). One of the issues linked to a disabling of participation in school-based curriculum development work appears to be the failure of the system to resource curriculum activities in schools (Committee for the Review of C2005, 2000; Czerniwicz *et. al.*, 2000) with appropriate learning support materials for teachers to draw on in designing relevant learning programmes. Added to this has been professional development processes that were often disembedded from the context of schooling and the real challenges teachers face in classrooms (Committee for the Review of C2005, 2000; Janse van Rensburg & Mhoney, 2000).

Evidence from the NEEP-GET pilot research and some case studies in the Learning for Sustainability Pilot project indicate that learning support materials and clear planning frameworks combine to provide the capital for processes of grounded inter-subjective curriculum development, interpretation and implementation practice amongst learners and educators (including teachers and curriculum support staff) in schools (Lotz-Sisitka & Raven, 2001). Findings from the Learning for Sustainability project indicate that clear curriculum and assessment frameworks are also an important dimension of enabling grounded inter-subjective talk-in-context and contextually significant curriculum work amongst teachers and learners (Lotz-Sisitka & Janse van Rensburg, 2000). Finding meaningful ways of enabling teacher participation in processes of interpreting curriculum policies in the context of school-based learning programme development is likely to continue to be a challenge for the education system (DoE, 2002).

Visible patterning

What we see in documents

It is not only how the curriculum is developed that is significant, but also the models and approaches that frame the curriculum, and how these are presented and interpreted at various levels in the system. I turn now to some of the tensions and challenges in the visible patterning efforts, notably the curriculum model, associated documents and assessment frameworks.

Models and approaches

The 1995 Environmental Education Policy Initiative (EEPI) proposed a range of possible approaches reflecting international trends to guide the inclusion of environmental education in defining a national curriculum. Debates on how environmental education should be included in the curriculum have been raging since the early 1970s (see Gough, 1997), and the consensus has been towards defining environmental education primarily as a cross-curricular concept.

The EEPI policy options include the following possible approaches to integrating environment into the curriculum:

- local, problem-solving curriculum action;
- integration within subjects;
- specialist/distinct courses (e.g. environmental studies);
- components within subjects.

These proposals were based predominantly on a subject/discipline model of curricula. Thus an early tension emerged for the environmental education community when the national curriculum development process started in earnest in 1996 with the release of a *Curriculum Framework Document* (DoE, 1996). The EEPI policy options appeared to have little value, as the boundaries of curriculum development processes had shifted to an outcomes-based orientation, with learning areas and outcomes, and not subject, as curriculum model. This required urgent re-orientation of environmental education contributions to curriculum debates, leading to the establishment of the EECI (an extension of the EEPI with a curriculum initiative focus, in early 1996). Jansen (1999) comments critically on the unexpected change in curriculum models being discussed in the Department of Education (DoE), and on the way in which the decision to adopt an outcomes-based education (OBE) framework was made without warning. Kraak (1999) provides some background to these decisions in his description of competing discourses shaping educational transformation in the early 1990s. These included a people's education discourse (characterising early resistance initiatives in the late 1980s), an integrationist discourse (featured in the ANC policy on education and training in 1994), and an outcomes-based discourse (arising in the labour movement's proposals for the re-orientation of education and training, and visible in the Department of Education's Curriculum Policy Framework for the first time in 1996). It is in the context of these competing discourses that environmental educators were attempting to formulate policy options, with earlier efforts clearly aligned with the ANC policy framework, and latter attempts aligning with OBE discourse.

Jansen (1999:7) describes the development of the OBE curriculum in the Department of Education as being characterised by an "underdevelopment of integration and competency propositions within the schools sector". He asserts that the "distraction of education development because of crisis conditions in schools and universities did not augur well for the development of curriculum policy in schools" and that the sudden emergence of OBE discourse in education led to a number of problems, including:

- a discourse on the curriculum which was foreign to teachers;
- confusion between two OBE discourses – one which developed in schools and

one which developed in vocational training;

- meaningless distinctions between traditional, transitional and transformational OBE models, adopted by the DoE from an American expert, Bill Spady;
- complex, voluminous terminology used to describe OBE, introducing more than 100 new words into the curriculum landscape (Jansen, 1999:7) – this latter problem was later described as an over-designing of the curriculum which was high on integration and low on progression (Committee for the Review of C2005, 2000).

Challenges rather than frameworks

Thus, the unfamiliar curriculum context provided wide-ranging challenges to environmental educators. In this uncertain context, it was difficult to adopt a particular model or approach (as outlined in the EEPI policy options document). It was more or less impossible to prescribe content or design old-style syllabi. EECI members were therefore faced with a set of challenges that were relatively undefined at the time. They were articulated by an EEPI working group in 1993 (O'Donoghue, 1993), and proved useful in guiding EECI interactions. They included:

- challenges to engage proactively with process and structural reform in curriculum policy;
- the need to clarify understanding of education, environment, development and environmental issues within policy and curriculum reform processes;
- the development of curriculum initiatives which enable teachers to exercise choice, to take responsibility and to engage in creative, challenging action to solve local environmental problems;
- the need to address, through education, global and local needs for environmental problem-solving, social justice and change towards sustainable living (O'Donoghue, 1993).

Contributing to curriculum policy frameworks and learning outcomes

In addition to initiatives to engage proactively with process and structural reform in curriculum policy were contributions environmental educators made in relation to the curriculum policy itself. These were not simply linked to defining content or methods in the form of contributions to curriculum documents, but instead involved a significant re-thinking of environmental education process in relation to emerging curriculum models and approaches.

An early challenge was to articulate an environmental focus in relation to the critical cross-field outcomes approved in 1995 by the South African Qualifications Authority to guide all qualifications and curriculum development activities. At a public meeting in early 1996, the EECI produced the first of a series of enabling documents (Janse van Rensburg & van Harmelen, 1996), which was presented at the first large-scale curriculum development gathering in Pretoria, to design C2005. This intervention from the EECI led to the defining of a number of environmentally-oriented outcomes in the eight learning areas, for example: "Learners will be able to make sound judgements about the management and utilisation of resources"; and "Learners will be able to address social and environmental issues in order to promote development and social justice" (DoE, 1997b:HSS-2). These outcomes were quite broad and open-ended, and designed to enable local, critical contextual action in school-community contexts.

However, once a complex assessment framework (detailing assessment criteria, performance indicators and expected levels of performance) had been defined for each outcome, the localised, critical and contextual intention of the outcomes changed, leading to difficulties in interpreting these learning outcomes in school and community contexts. At the time, there were many debates about the instrumentalist nature of OBE, and outcomes-based curricula (see for example McKernan, 1993; Jansen, 1997). These debates included discussion about the dangers of assessment-driven curriculum interpretation. The DoE resolved this tension by articulating a transformative approach to OBE (influenced by Bill Spady as mentioned above). Unconvinced of Spady's theories, and mindful of Jansen's (1997) critique of OBE, some members of the environmental education community attempted to resolve this tension in the field with strategies that encouraged educators to develop learning programme activities that were learner-centred, while keeping the curriculum outcomes in mind (School Environmental Policy Pack, 1998; Lotz, 1999a; Bjerg-Hansen & Sguazzin, 2000; Lotz-Sisitka & Raven, 2001).

This development led to a shift towards school-based curriculum development activity, which became a focus of the EECI[2] and the Learning for Sustainability Pilot Project.[3] This shift enabled more open-ended, contextually responsive learning processes in schools (Lotz & Olivier, 1998). However, the potential of this shift never really materialised, as school-based curriculum development interactions were often limited to curriculum design activities in teacher professional development settings. Very little environmental education curriculum development made its way into schools and classrooms, and the professional development settings (often based on workshops and linear cascade approaches) did not provide teachers with adequate opportunities to narrate changing curriculum practices. In an attempt to create opportunities for an applied, reflective focus for professional development, the Learning for Sustainability Project piloted a cluster-based spiral model for professional development (Janse van Rensburg & Mhoney, 2000), a model which is currently being applied in the NEEP-GET.

Where curriculum design and interpretation was attempted in classrooms, the changes were often superficial. One of the problems in early implementation attempts was the fact that little or no attention was paid to the assessment frameworks attached to the learning outcomes (mainly because of their complexity). This often contributed to superficial interpretations of the curriculum (Lotz-Sisitka & Olivier, 2000), as teachers found it difficult to establish appropriate levels of scope and depth for learner groups. These kinds of problems led in many cases to poor quality teaching and learning opportunities, and a marginalising of assessment (Lotz-Sisitka & Raven, 2001; see also Taylor & Vinjevold, 1999).

Problems associated with the implementation of the curriculum framework were, however, not only linked to difficulties in interpreting and mediating the policy documents. They were (and are still) linked to the challenges associated with social transformation.

Drawing on the works of Durkheim & Bernstein, Harley & Parker (1999:195) provide an insightful analysis of what they describe as a "mismatch between policy and society". They describe the way in which South African society was historically organised primarily on the basis of mechanical solidarity, while the transformation agenda (including the Constitution and C2005) reflect a desired but unfamiliar organisational framework of organic solidarity. Their analysis sheds light on the difficulties experienced by educators in localised interpretations of curricula. The strong participatory tendency in the fledgling democracy, fuelled by critical ideologies prominent in environmental education, failed to empower teachers. Instead, it burdened them with the expectation of developing learning programmes to respond to the local context and issues, and to interpret a complex curriculum framework, with little support (Committee for the Review of C2005, 2000). Factors contributing to these difficulties included the hybrid nature of the curriculum approach, new identity constructions, a strong integrationist discourse with weak knowledge classifications, blurring of boundaries, and weak curriculum framing. These factors lead to a "highly complex" new role for teachers as curriculum developers (Harley & Parker, 1999:191; see also Lotz & Olivier, 2000).

Across the curriculum or integral to the learning areas?

In addition to these problems came attempts to integrate environmental education into the curriculum. During a public meeting in 1997, the EECI produced a second enabling document which aimed to influence the way in which learning programmes were constructed. *Enabling Environmental Education as a Cross Curricular Concern in OBE Learning Programmes* (Janse van Rensburg & Lotz, 1998) illuminated an environmental focus for each learning area, and presented several models for learning programme development which illustrated how environmental issues could be integrated across the curriculum. These included an outcomes/thematic and issues-based approach to learning programme development.

The document was distributed and used at another national stakeholder meeting for the development of C2005, with the result that environment came to be described as one of six phase organisers or integration features of the curriculum. This led to wide-ranging initiatives to support teachers in designing corresponding school-based learning programmes (see for example Lotz, Tselane & Wagiet, 1998; Lotz & Olivier, 2000). Significant in the discourse of environmental education curriculum development in South Africa was the fact that environmental educators were no longer trying to get environmental education included in the curriculum, but were now faced with the challenge of interpreting environment in the context of the curriculum policy. Thus the focus in curriculum work shifted to environment as organiser of curriculum activities. This enabled teachers to consider an environmental focus in the context of the various learning areas. However, with the national review of the curriculum in 2000 resulting in a criticism that it was over-designed, phase organisers were dropped (Committee for the Review of C2005, 2000).

Thus, following the review of the curriculum pending in early 2001, and the emergence of a streamlined C2005 in 2002, one of the key issues to be addressed in the NEEP-GET pilot research process was the way in which environment would be conceptualised. As noted above, the dominant discourse in environmental education positioned environment as a cross-curricular issue. With the streamlining of the design features, seeing environment as an organiser was no longer an option, and environmental educators shifted their focus to reconceptualise the way in which environment was being interpreted in the learning areas. In the process, they could draw on the experience of school-based curriculum development and interpretations in the EECI and Learning for Sustainability Projects, in which teachers designed activities in response to particular environmental issues and risks, with outcomes in mind (using an integration-across-the-curriculum approach). Findings from the EECI, Learning for Sustainability Project and NEEP-GET pilot research indicated that many of the activities being planned in schools were not furthering the aims of the learning areas, nor were they enabling a deepening of knowledge and process skills in specific learning areas, in many cases leading to superficial or misinterpretations of the learning outcomes with associated superficial learning outcomes (Lotz-Sisitka & Olivier, 2000; Lotz-Sisitka & Raven, 2001). For example, environmental activities were designed for the natural sciences learning area, with little evidence of learners engaging with scientific skills.

The NEEP-GET Project led to a further enabling document, this time aimed at supporting groups designing the revised National Curriculum Statement (NCS). This document (NEEP-GET, 2001) clarified an environmental focus integral to each learning area. For example, learning outcomes in the technology learning area would emphasise the design and development choices and technological options available for sustainable lifestyles and improved environmental management. Learning outcomes in the natural science learning area would emphasise investigations and understanding of biodiversity and life-support systems, as well as risks and threats to life-support systems. The social sciences, on the other hand, would emphasise human-environment relationships (past and present), particularly those relating to landscape change and current development options, as well as to citizenship and sustainability issues (Lotz-Sisitka & Raven, 2001; NEEP-GET, 2001). NEEP-GET staff are currently working on a transitional mechanism to support educators in designing curriculum activities which emphasise the environmental focus integral to each learning area while enabling, where appropriate, integration around local issues. The NCS has also foregrounded a focus for environmental learning in the curriculum, emphasising the relationship between human rights, a healthy environment and social justice (DoE, 2002). This focus is consistent with the Constitution and sets a clear patterning framework for future environmental education processes (within the learning areas) in schools.

These developments represent a significant shift in environmental education curriculum work, particularly the uncoupling of the notion of environmental education as a thing in its own right to a foregrounding of environment in the curriculum. The abovementioned principle of the NCS adds further perspective to the focus on environment in the curriculum, by highlighting an environmental focus in each learning area, in a manner that is consistent with the human rights/social justice agenda of the curriculum in post-apartheid South Africa. These developments also illustrate the shift that has taken place: from seeing environment (and sustainable development) as a peripheral or marginal issue in the curriculum, to seeing it as an issue that has become a key focus in curriculum policy.

Policy and practice

Still mindful of the earlier EEPI position on policy and practice, and the experience gained in the abovementioned interpretation and implementation of curriculum policy in the EECI and Learning for Sustainability Projects, environmental educators are still grappling with ways of enabling environmental learning processes in schools. The early work of the EECI, which introduced a number of initiatives to support curriculum implementation in schools, is now being taken further in the context of the NEEP-GET. Amongst the practical school-based curriculum initiatives is the development of a *School Environmental Policy Pack* (1998, learning support materials to support environmental management plans in schools, and an auditing of curriculum activities and school resources such as water and energy).

Many environmental education service providers are developing learning support materials to support environmental learning processes in schools, but these are on the whole uncoordinated and uneven in quality and focus. The NEEP-GET aims to provide orientation to these partners to develop learning support materials that are more consistent with curriculum policy. A key finding in the Learning for Sustainability Pilot Project (Lotz-Sisitka & Olivier, 2000) and the NEEP-GET pilot research was a clearer understanding of the role of learning support materials in OBE. This is currently being explored in the context of NEEP-GET in line with DoE policy on resource-based learning and in the context of the difficulties associated with resourcing under-resourced schools in South Africa (Czerniewicz *et al.*, 2000). In particular, the NEEP-GET is giving attention to the notion of learning support materials, or materials that are prepared in ways that structure and support environmental learning (Czerniewicz *et al.*, 2000:20). The DoE's *Guidelines for the Development of Learning Support Materials* (1998) indicate that learning support material should be viewed as an integral part of curriculum development, emphasising the important role that a careful consideration of learning support materials and resource-based learning approaches play in enabling curriculum change.

A further dimension of the struggle to consider policy and practice were efforts in the Learning for Sustainability Pilot Project to clarify the government's environmental policies in relation to education. In particular, the project focused on the notion of sustainability, aiming to clarify this in the context of curriculum work (Janse van Rensburg & Du Toit, 2000). In line with global efforts, the Learning for Sustainability Project aimed to link ecosystem conservation with socio-economic development, and in the process encountered many complexities. As a departure point the project focused mainly on contextual environmental issues. The difficulties the project faced in introducing teachers to context-specific environmental issues and a discourse of sustainability included teachers' limited knowledge of local issues, and the fact that they lived and worked in contexts where a much stronger discourse of economic development does not clearly support or link with environmental objectives (Janse van Rensburg & du Toit, 2000; Lotz-Sisitka & Janse van Rensburg, 2000). While environment had been introduced in the policy milieu, it was not complemented by visible government action on the ground, and teachers found it difficult to see actual options for sustainable development in a local context. This illustrates that even if environment is in the curriculum, there are many complex factors associated with its local interpretation. The disjuncture between visible policy and visible practice is likely to continue to be a challenge for curriculum work in future.

What is in and what is out?

In spite of such wide-ranging and complex engagements, a curriculum development process is, by its very nature, perspectival, and can only represent particular issues. It took a travelling text worker to illuminate perceived silences in environmental education curriculum development interactions (Gough, 1998). In his comments on EECI perspectives informing curriculum debates at the time, Australian curriculum theorist Noel Gough also noted that the EECI perspectives were reflective of dominant neo-liberal economic discourses (which had been critiqued as contributing to environmental issues and social inequalities). Of course, for curriculum developers involved in the processes at the time, these retrospective interpretations were clinically clear, but never that clear when the issues were being deliberated. Simply arguing for environmental perspectives in an economic and management sciences curriculum was a noisy process. I remember hot contestations of particular economic models being proposed as guiding frameworks for the economic and management sciences curriculum, and equally heated debates about whether South Africa's macro-economic framework, GEAR, was appropriate for development in the country. More recently, in designing the NCS, debates related to the economic and management sciences curriculum have been similar, particularly on the way in which the curriculum reflected support for globalising neo-liberal economic models. Enabling the space for critical (environmental) interpretations in these curriculum development processes was particularly challenging at the time, and will no doubt continue to be so in the future, illustrating the (often) contested nature of the patterns inscribed in curriculum policies.

Process patterning

The shaping of learning interactions

A further dimension of curriculum development work is the shaping of learning interactions through the way in which the curriculum is designed. Ideas about teaching and learning processes have provided a rich arena for debate, struggle and contestation in environmental education over the past few years. Bauman (2001:123), commenting on the work of Mead and Bateson, notes that they assigned the "primal and decisive role in the process of teaching and learning to the social context, and the mode in which the message is conveyed, rather than to the contents of the instruction". Drawing on Bateson's work, Bauman (2001:124) introduces the importance of what he calls "deutro learning", or the way in which learners learn, and the important relationship between learning how to learn (learning processes) and what one learns (the visible curriculum).

Old versus new

Early attempts at grappling not only with what goes into the curriculum but also with how learning takes place often resulted in a tendency to oppose past dominant learning theories and associated methods with more contemporary learning theories and methods. This trend was evident in a number of DoE documents which presented changes in learning processes in the form of simplistic tables contrasting the new (learner-centred education) with the old (teacher-centred education). In the former approach, the learner is described as being active, while the educator is to take on the role of facilitator.

In professional development settings, historical framings of changes in learning theories reflected similar shifts, albeit with more detail. The results often comprised a linear, historical narrative framed within modernist time-space configurations (Popkewitz, 1999). In this narrative, early behaviourist theories are essentially rejected in favour of more popular constructivist theories (particularly social constructivism) thus presenting something of a grand narrative to shape environmental education processes (see for example the Rhodes University/Gold Fields Participatory Course Materials, 1995 to 2000).

These perspectives influenced some of the curriculum development initiatives described above, particularly the Learning for Sustainability Project. They resonated with the learner-centred focus of the OBE curriculum framework at the time, and with early articulations of the transformation of learning processes by the DoE, in the form of an oppositional discourse (constructivism versus behaviourism). It is significant to note that it was only in 2000 that the DoE actually articulated a theoretical framework for OBE, which included fairly superficial descriptions (and misinterpretations) of constructivism as a theoretical framework for teaching and learning (DoE, 2000; Moll, 2001; 2002).

Early curriculum policy documents all proposed a paradigm shift from a traditional aims-and-objectives approach to that of an outcomes-based one. The implications of this "paradigm shift" for learning processes were not clearly articulated. In 1997, a conference on OBE caused a stir in educational circles when the OBE curriculum was critiqued by prominent academics for a lack of attention to the implications it would have on learning processes. Parker (1997), for example, noted that one of the dangers lay in the technicising of OBE and the potential for outcomes to be rote-learned (thus perpetuating some of the problems of apartheid education). Early DoE attempts to create a theoretical framework for OBE (drawing on the work of Spady) were critiqued for limited and limiting views of curriculum and learning (Committee for the Review of C2005, 2000:11).

Learner-centred emptiness

Learner-centred education was first introduced into South Africa's curriculum policy discourse in the early days of People's Education (Kraak, 1999). The rhetoric surrounding this approach was included in OBE discourse contributing to the hybrid model noted before. Many environmental education curriculum initiatives (including some EECI activities, and the Learning for Sustainability Project) linked learner-centred pedagogy and constructivism, often drawing on the oppositional discourse proposed by the DoE to justify a focus on constructivism (Janse van Rensburg & Lotz-Sisitka, 2000). With its emphasis on context and the social in environmental education, social constructivism (Gergen, 2001; Janse van Rensburg & Lotz-Sisitka, 2000) became a popular educational theory shaping environmental education processes. However, the results of imposing these discourses on curriculum development processes are superficial, naïve and limited interpretations of both learner-centred education and constructivist learning theories (Lotz-Sisitka & Olivier, 2000; Janse van Rensburg & Lotz-Sisitka, 2000; Taylor & Vinjevold, 1999). This also impacted on the role perceptions of teachers, who were challenged to see themselves in their new roles as facilitators, rather than as dispersers of knowledge. Taking note of these problems,[4] the DoE changed its discourse: teachers were no longer referred to as facilitators, but were now mediators.

The Learning for Sustainability Project grappled with these developments during its piloting processes. Teachers engaged in the project learned to use concepts such as constructivism and prior learning, but in most cases they equated these concepts with activity-based learning and group work. The notion of learners doing things together in groups was seen to be constructing knowledge in learner-centred ways. Closer probing of these issues indicated that teachers were modelling surface forms of learner-centred activities without understanding the learning theories underlying them and without using them to mediate learners' engagement with substantive knowledge and skills (Janse van Rensburg & Lotz-Sisitka, 2000).

Furthermore, Taylor & Vinjevold (1999:20) note that the epistemological framework of the progressive consensus guiding South Africa's educational reform is constructivist "... despite the fact that it constitutes one of the most controversial and difficult visions to implement in education today".

Open processes of active learning through OBE

Struggling with these systemic-conceptual problems, environmental educators took their lead from the emphasis on active learning through OBE in the Minister of Education's Tirisano Strategy (Asmal, 1999). Soon after the establishment of the NEEP-GET (a Ministerial initiative), the emphasis on active learning through OBE featured in the first NEEP publication, *EnviroDays 2000* (DoE and EEASA, 2000). This approach was piloted in the NEEP-GET research, in which an open-process framework for guiding active-learning processes was developed and tested in over 70 schools around the country. It was linked to developing NEEP-GET perspectives which recognised that environmental issues and risks are contested and without clear-cut solutions, and that as a result, environmental education processes and learning outcomes needed to be open-ended (NEEP-GET, 2001). The document also emphasises that "... enabling learners to address environmental issues requires knowledge and skills best developed through active learning, critical thinking, involvement in real issues and encounters in [their] immediate environment" (Lotz-Sisitka & Raven, 2001:94), thus drawing attention to the importance of the pedagogical challenges (Dreyfus, Wals & van Weelie, 1999) in environmental education curriculum work.

O'Donoghue (2001:5) articulated these challenges as being linked to "contextual social processes of cultural induction" and "critical reflexive re-orientation within open-ended educational processes aimed at sustaining human interactions in healthy, just and equitable environments".

Thus the NEEP-GET piloted an open-ended process framework to provide orientation to teachers in planning education processes of active learning and change. These included a mix of processes that enable learners to find information about issues, explore these through experiences in the environment, and take action for a better world. O'Donoghue (2001) notes that these processes have the potential to contribute to meaningful environmental learning, and to better environmental management and more sustainable lifestyle choices. The NEEP-GET pilot research indicated that processes which mobilise prior knowledge (including indigenous knowledge) and experience are also significant in environmental learning (see O'Donoghue & Neluvhalani, this volume). This open-process framework thus provides a framework for curriculum interactions which respond to both the environmental and the pedagogical challenges of environmental education in local contexts (Lotz-Sisitka & Raven, 2001).

Socially critical steering ideas

A further dimension of process patterning that has gained considerable ground in environmental education curriculum work in South Africa is that of socially critical pedagogical approaches (as articulated by Fien, 1993; Greenall Gough & Robottom, 1993). The potential and the limitations of these steering ideas, were highlighted in the Learning for Sustainability Project. According to Janse van Rensburg & Du Toit (2000:39), important dimensions of socially critical perspectives include:

- giving attention to the socio-political and economic dimensions of environmental issues. The potential thus lies in encouraging teachers and learners to deepen and challenge limited and superficial understandings;
- the potential of socially critical pedagogies to focus on active learning through action to address environmental issues (see also Greenall Gough & Robottom, 1993; Gough, 1997), thus enabling school-community links and community action;
- the importance of foregrounding value positions and taken-for-granted perspectives, which can facilitate critical, contextual re-orientation, and a re-thinking of ideological assumptions as outlined by O'Donoghue (2001).

Janse van Rensburg and Du Toit (2000:39) also note that the weaknesses of socially critical pedagogies include:

- the danger of the educational process becoming deterministic, particularly if the educator holds too strongly to critical structuralist explanations of issues;
- the danger of being over-optimistic about the potential of teachers and learners to address environmental issues. Education (for sustainability) has limited potential to address these issues directly, not only because of the limited place of environmental agency and action in schools, but also because of broader systems and discourses, which education more often than not reflects rather than challenges.

These observations are echoed in some of the earlier points in this chapter about curriculum as patterning; collectively they illuminate the significant tensions and challenges associated with environmental education in processes of social transformation. In conclusion, I focus again on this aspect.

Towards reflexive patterning in curriculum development work

The most recently published curriculum policy document (released as this paper was written) states unequivocally that "teachers will be responsible for learning programme development" (DoE, 2002:12). In reflecting on this statement, in relation to the narrative above, I tentatively and "summatively" comment on some of the key challenges for ongoing environmental education curriculum work. These may include:

- a re-orientation of "patterns of participation", in which we locate these patterns of participation in contexts of school-based curriculum deliberations with teachers;
- a re-orientation of professional development processes to support teachers' curriculum development work, in ways that enable them to narrate curriculum

change practices, with responsive attention being given to school-based curriculum deliberations and the resourcing of environmental learning in the various learning areas, in the context of local issues and risks;

- a re-orientation of "visible patterning", in which the "visible curriculum work" is constituted not in policy documents and other curriculum artefacts, but by evidence of environmental learning in schools;
- a re-orientation of "process patterning" in ways that refrain from proposing "grand narratives" of method and learning theory to educators (which often tend to become rhetorical) – approaches and frameworks that enable educators to make diverse methodological choices that are appropriate to context, learning area and learners would seem to be more useful;
- development of a more in-depth and critical understanding of the pedagogical, epistemological and reflexive challenges associated with emerging patterns of environmental risk in curriculum work. How do we engage open processes of "critical reflexive re-orientation" (O'Donoghue, 2001) in ways that will enable sustained human interactions (and developments) in healthy, just and equitable environments?

In "ending" this narrative, I wish to tentatively open up aspects of this latter challenge for further consideration and debate.

In commenting on the trends and challenges associated with "process patterning" in curriculum development work, I drew attention to the fact that environmental educators have increasingly turned their attention to the way in which learners learn. One such process has involved a consideration of DoE suggestions (and theory) on learning processes in relation to OBE, educational transformation and environmental learning processes. I would argue that a further dimension of curriculum development work involves paying attention to a "reflexive patterning" process in environmental learning.

As noted in the backdrop to this chapter, Bauman draws our attention to the limitations on reflexion in enabling social transformation. He comments on the limitations of current reflexive practices, noting that we may well require a reflexivity that enables us to consider not only our life political choices, but also the nature of our critique, in relation to the agenda set for our life political choices. Bauman (2000:23) notes that increased individualising of contemporary society is "...rendering our critique toothless". He notes that contemporary society has invented ways of "...accommodating critical thought and action, while remaining immune to the consequences of that accommodation". What, therefore, would the purpose be of encouraging learners to "critically engage with environmental issues and risks" in curriculum processes, if their critique is likely to be rendered meaningless through society's ability to accommodate critical thought and action, while remaining "immune" to it?

Critical thought and action, uncertainty and ambivalence

A key challenge for environmental education curriculum work that wants to engage social transformation would therefore seem to be located in a consideration of the nature and purpose of critical thought and action. Findings in the Learning for Sustainability Project (Janse van Rensburg & Du Toit, 2000:31) indicate that perspectives of environmental issues were often characterised by "narrow interpretations" of the issues, when causes of environmental issues were, for example, attributed to ignorance of individuals and other deficiencies. These "narrow interpretations" showed very little evidence of systemic or historical analyses in teachers' understandings of environmental issues, and associated learning opportunities provided to learners. Janse van Rensburg & Du Toit (2000:32) note further that limiting critical engagements to a focus on constructing meaning in local context may, in fact, contribute to a narrowing of the possibilities for engaging critique of issues. They argue that teachers need more information, access to information and critical perspectives on information about a wide range of issues for meaningful critical engagement with contextual issues, and in order to teach effectively in an outcomes-based framework. The NEEP-GET pilot research indicated that not only is information an important dimension of critical, reflexive work in environmental education, but *open processes* of enabling local engagement with issues and risks through investigations, action-taking and reporting are also important.

Popkewitz (1999) draws attention to the changing terrain of the critical project in education. Drawing on the work of Foucault, he points to changing views of power-knowledge relationships, and ways in which knowledge and pedagogy may "discipline and regulate" people. The challenge in environmental education curriculum work may thus extend from clarifying orientations to critical work, to engaging critically and reflexively with governing principles that organise individual action and participation in school-community interactions. Bauman (2001:139) notes that the "uncertain and ambivalent" character of social life, evident in the complexity, diversity and often limited solutions associated with environmental issues and risks (Lotz, 1999b; Lotz & Robottom, 1998), requires a "fortifying of critical and self-critical faculties", and a "capacity for changing the frames". He notes that these skills can hardly be developed in the context of educational processes that are designed to control and direct learning, thus posing critical curriculum questions for ongoing consideration.

Knowledge, unawareness and inter-epistemological discourse

The Learning for Sustainability Pilot Project and the NEEP-GET pilot research highlighted the importance of considering ways of enabling learners to engage critically and reflexively with environmental issues and risks. These projects have also drawn attention to the epistemological, political, pedagogical and practical challenges associated with these processes. Odora (1994:101) comments insightfully on the epistemological and political dimensions of curriculum transformation in Africa. She sees African "environmental knowledges" (those knowledges that have thus far been subjugated and rendered irrelevant by a scientism characteristic of modern Western epistemology and education systems) as an important facet of educational transformation in Africa. She notes that there is a need for careful attention to the "insurrection of subjugated knowledges", including knowledges that have been "buried under the yoke of formal systematisation". Commenting critically on the tendency to systematise knowledge in formal education systems, she notes a trend in which environmental knowledges in Africa are becoming limited to an understanding of environmental problems in ecological, physical and technical terms, and where the problem of the environment is constantly defined as loss of soil, disappearance of forests, extinction of species, spread of deserts, pollution of waterways, etc. She also observed that "the villains in the piece are simplistically defined" (Odora, 1994:101).

Beck (1992), in his seminal work *Risk Society* draws our attention to the fact that patterns of environmental risk are complex, and require a "reflexive society" and inter-epistemological dialogue. More recently in his analysis *Knowledge or Unawareness?* he alerts us to the fact that reflexivity involves *reflection* (knowledge of what is known) and *reflex* (the effect or preventative effect of *non*-knowing), noting that there are various forms and constructions not just of knowledge, but also of unawareness (what we don't know, and why we don't know) (Beck, 1999:109). What do/can learners know about environmental issues and risks (how do we mobilise prior knowledge and experience)? How can they find out? What don't we know, and why don't we know? These may be some of the more obvious curriculum questions that would enable an epistemological and political reflexivity in environmental education curriculum deliberations between educators and learners.

An open-ended closure

In this chapter, I have reported on a range of patterns and challenges associated with environmental education curriculum development in South Africa over the past ten years. These have been contextualised within the curriculum development processes taking place in post-apartheid South Africa. Drawing on a backdrop of social theory, curriculum theory and international developments in environmental education, I have tried to illustrate that the challenges associated with curriculum patterning processes are broad-ranging, complex and multi-dimensional.

As the patterns of environmental risk (and associated responses) appear to become more complex and ambivalent in the context of increased globalisation and international contestation (evident in debates surrounding the World Summit on Sustainable Development), the associated educational response appears to require ongoing (critical) reflexivity and re-orientation. In attempts to "mainstream" environmental education processes in the formal education system in South Africa we have learned much about the challenges associated with the educational response to environmental issues and risks, particularly as these relate to curriculum development work in a context of socio-political transformation. This review does not provide definitive orientation to ongoing curriculum work, but rather leaves us with an open-ended challenge to continue engaging educationally with patterns of environmental risk in the context of formal education curriculum policy and practice.

Bauman's (2001:139) comment on the challenges for educational philosophy and theory provide a fitting open-ended closure to this paper:

> ...educational philosophy and theory face the unfamiliar and challenging task of theorising a formative process which is not guided from the start by the target form designed in advance; modelling without the model to be arrived at in the end being known or clearly visualised; a process which can at best adumbrate, never enforce, its results and which builds that limitation into its own structure; in short, an *open-ended process, concerned more with remaining open-ended than with any specific product* [my emphasis] and fearing all premature closure more than it shuns the prospect of staying forever inconclusive. This is perhaps the greatest challenge which the philosophers of education, together with the rest of their philosophical colleagues, have encountered in the modern history of their discipline.

Acknowledgements

In writing this paper, I acknowledge the many environmental educators who have contributed to the processes reviewed here.

This paper was produced as a contribution to the Rhodes University Environmental Education Unit's "Environmental Learning and Curriculum" Research Programme funded by the National Research Foundation (NRF).

Endnotes

1 See, for example the President's Education Initiative (Taylor & Vinjevold, 1999); the Curriculum Review (Committee for the Review of C2005, 2000), the Environmental Education Curriculum Initiative (Lotz, Tselane & Wagiet, 1998) and the Learning for Sustainability Project (Janse van Rensburg & Lotz-Sisitka, 2000; and Lotz, 1999).

2 See for example Lotz, Tselane & Wagiet, 1998.

3 See for example Lotz, 1999; Lotz & Olivier, 1998; 2000; Du Toit & Sguazzin, 2000; Bjerg-Hansen & Sguazzin, 2000.

4 Recent work by Moll (2001; 2002) and the President's Education Initiative research project (Taylor and Vinjevold, 1999), are providing useful perspectives on some of the problems associated with superficial, naïve and limited interpretations of constructivist learning theories in South African curriculum discourse.

CHAPTER TEN

Indigenous Knowledge and the School Curriculum:

A Review of Developing Methods and Methodological Perspectives

Rob O'Donoghue and Edgar Neluvhalani, South Africa

Background

This chapter explores developing methods and methodological perspectives for mobilising indigenous knowledge within environmental learning activities in the school curriculum. Methods are ways of fostering learning, and methodology provides explanation and justification for approaching teaching and learning in particular ways.

Rob O'Donoghue drafted the chapter in an attempt to pull together a decade of somewhat *ad hoc* cooperative resource development work through Share-Net.[1] Edgar Neluvhalani's comments on the text, while he was editing a UNESCO module on indigenous knowledge (UNESCO, 2001) and writing a Ph.D. research proposal (Neluvhalani, 2002), developed into this collaborative review of methods and methodologies. The review was undertaken to feed into a developing research programme at the Rhodes University Environmental Education Unit and as a contribution to a seminar on Endogenous Knowledge convened at Pretoria University.[2]

Educational perspectives on indigenous knowledge processes are reviewed as these emerged in research and environmental education resource development work over a period of about ten years. In examining the evidence and reflecting on the experiences, we try to differentiate methods and to clarify methodological perspectives for current school curriculum settings. The purpose of the exercise is to document evidence and to mobilise a conceptual capital for making more of sustaining patterns of indigenous knowing in environmental learning activities in the school curriculum.

Knowledge and knowing

In approaching the challenge of reviewing developing methods and methodology, we often use the term "indigenous knowing" in preference to the more common notion of "indigenous knowledge". This is done to avoid treating knowledge as an objective commodity outside the socio-historical contexts of human meaning-making interaction.

A knowing-in-context perspective can help one to avoid locating knowledge construction and mediation processes within an oppositional logic. Dialectic perspectives separate out and oppose indigenous/non-indigenous and Western/African scientific knowledge, for example. Knowledge differentiated in/as abstracted systems that are seen as exclusive and opposing can narrow and limit an engagement with the dynamic capital of knowledge sedimented amongst peoples interacting in the socio-historical contexts of southern Africa.

The idea that knowledge is a commodity people possess and that it is located in an indigenous knowledge system also denies an appreciation that much of our human cultural capital of knowing in, and of, the everyday, is not always expressed as explicit matters of fact. We all have a social habitus (Bourdieu in Bourdieu & Passeron, 1977), an orientating capital of perspectives in and for everyday life. This is a relatively blind personal and cultural capital of habitual knowing that one finds encapsulated in many of the certainties we experience as we see, feel and do things in particular ways. Within this conventional wisdom, we often feel and know what matters for life in our given context, more often than not without the need to articulate any explanatory facts of the matter – the "why".

For the purposes of this chapter, situated knowledge that is commonly sought and referred to as indigenous knowledge is taken to be a diverse capital of sustaining, inter-generational knowing in, of and for life in a particular socio-ecological setting. To examine, clarify and inform methodology and methods for mobilising this capital in the school curriculum, we first review historical evidence and more recent case experiences from our study and work that have shed light on and come to mobilise a rich capital of indigenous knowing in southern Africa. We then examine environmental education resource materials and methods in cases of environmental learning in school curriculum contexts.

The review documents a chronological South African story of emerging methods and methodological perspectives for mobilising indigenous knowing in and for contexts of environmental learning in schools. Historical evidence of indigenous knowledge and practical cases of interpretative re-appropriation are probed. Methodological insights are reviewed within case histories where indigenous knowledge was mobilised within interactive, meaning-making struggles around everyday practices and in practical reconstructive work on an indigenous technology. The curriculum application of interpretative texts is then critically examined, along with a review of a variety of developing materials and methods. These are examined within some of the key dimensions of the active learning framework used by the National Environmental Education Programme to foster environmental learning in the new outcomes-based education curriculum for South African schools.

Introduction

The story began alongside the 1992 Earth Summit in Rio de Janeiro and at a time of significant socio-political change in South Africa. The principles for Environmental Education for Equitable and Sustainable Societies, adopted by the International NGO Forum at the Earth Summit, were a mobilising influence in developing work on indigenous knowledge processes among South African environmental educators. Notable are principles 7 and 9 (ICAE, 1993).

Principle 7 maintains that:

> Environmental education must recover, recognise, respect, reflect and utilise indigenous history and local cultures, as well as promote cultural, linguistic and ecological diversity. This implies acknowledging the historical perspective of native peoples as a way to change ethnocentric approaches, as well as encouraging bilingual education.

While principle 9 advocates that:

> Environmental education values all different forms of knowledge. Knowledge is diverse, cumulative and socially produced and should not be patented or monopolised.

This set of guidelines resonated with the critical, democratic and transformational values and ideals promoted in a fledgling South African democracy. Despite this, as well as a broader international focus on indigenous knowledge in the environmental arena, local initiatives were slow in gathering momentum. An unpublished working document probed some of the socio-historical blockages in early and emerging South African narratives (O'Donoghue, 1994).[3] It was only as early environmental education materials (Manqele, 1994) came into wider use and case-study research (Masuku, 1999) began to open up the concern that curriculum perspectives on indigenous knowledge began to develop.

While the last few years have seen an explosion of interest in the wake of President Thabo Mbeki's call for an African Renaissance, and while there has been much activity around indigenous knowledge in the development, political, scientific, academic and cultural arenas, not much that is tangible has materialised and, as yet, little has been translated into curriculum perspectives and learning support materials.

Appropriating oversight, past and present

The colonial histories of southern Africa are full of cases where indigenous peoples and practices were regarded as barbaric or uncivilised against the more enlightened precepts of colonists. In contrast to and alongside these perspectives and the appropriating politics of colonising prejudice, one also finds cases where outsiders developed insights on, and a respect for, the common-sense wisdom of indigenous peoples. It is hoped that a chronological review of illustrative, historical case evidence of indigenous knowledge processes, as well as experiences of the use of this and other materials in curriculum contexts, will help us address the methodological challenges of mobilising indigenous knowledge in schools today.

For example, historical evidence illustrates how insightful knowledge was often overlooked and marginalised in the past. Notable in African contexts is how situated knowing was often appropriated and transformed within outsider ways of seeing things. An appreciation of these processes allows one to probe ways in which knowledge became sequestrated from the everyday with the advent of technological expertise and the modern state (Bauman, 1993:196). With illustrative historical evidence, one is able to challenge some of the naïve oppositional assumptions referred to above, and much of the ethnocentrism that can bedevil an open-ended clarifying of methodological perspectives on indigenous knowledge.

Appropriation (Giddens, 1991) is a useful concept for probing the politics of knowledge creation. It allows one to engage with processes of oversight, transformation and marginalisation through which much earlier contextual knowing was lost and changed in the politics of human meaning-making. An examination of historical case evidence within the chronology of the developing story can provide useful insights for educators to engage the capital and depth of indigenous knowing apparent in many ways of old.

Knowing what to do in encounters with wildlife

Historical narratives that shed light on the loss, transformation and appropriation of indigenous knowledge are to be found in texts of earlier times. For example, the Colenso diary *Bringing Forth Light* (Edgecombe, 1982) is an interesting case. The historical records allow one to juxtapose what the bishop and his students from the mission school documented as daily happenings on a trek into Zululand to visit King Mpande in 1859. On their way back, the bishop writes of taking the wrong road and of encountering buffalo. In reporting this event and his observations on how the local people rallied to hunt and drive away this wild animal, a threat to their cattle, he makes no mention of how his young Zulu students started a fire in response to the threat.

The student Magema Fuse, on the other hand, reports how they took some dung and grass to "make it smoke upon the oxen". He also writes about the bishop questioning this activity:

> Sobantu [Bishop Colenso] enquired and said, What is this? What are you doing it for? They said, Why this now, people do it that so, when cattle arrive at the place of others [wildlife], they may not come to take their sickness (Edgecombe, 1982:169, notes in brackets added).

Neither the bishop nor any of the students explains the common-sense wisdom in this indigenous practice. In 1859, the colonists saw the tsetse fly as a finger post beyond which they and their animals were at risk. However, the colonial administration did not yet explicitly know that the *Nagana* (sleeping sickness) was transferred to cattle from large wild animals like the buffalo by the tsetse fly. It was certainly not yet known that the tsetse fly, unlike other flies, hunts mainly by sight.

At the smell of smoke, not unlike ticks that drop off grass to hide in cracks in the soil, it flies off to the shelter of the moist shaded glades of river valleys, its breeding grounds.

It is also unlikely that the young Zulu men had these explicit insights at that time. They did, however, have an inter-generational common-sense wisdom that enabled them to know what was to be done, "when cattle arrive at the place of the others".

It is also somewhat ironic that, just before the turn of the century, a scientist explicitly established the relationship between cattle, wildlife and the tsetse fly to describe the transmission of the *Nagana*. However, the "new" knowledge was sequestrated from daily life and appropriated into the colonial administration to be set up against earlier indigenous myths. Now confident of knowing more and better than "the Other" (the local populace), it successively took up the role of educating indigenous people. The administration seldom appears to have had much respect for, and certainly overlooked much of, the knowledge that indigenous peoples had in common-sense ways of doing things in the contexts of their everyday lives.[4] These are similar to processes during state formation in Europe and elsewhere. With the advent of the modern state, much of an earlier and local capital of knowing was assigned to the margins as myth and superstition (Bauman, 1993).

Still overlooked and in the margins today

The overlooking and marginalising of indigenous wisdom in local ways of knowing and doing things is not only a thing of the past. Two more recent cases in our experience were:

- When interviewed by young learners about past water use, an elderly man told an exaggerated story about how the rivers used to be bigger and fuller in earlier times. He also described how, despite there being ample water in the rivers, people collected their water needs from pools of rain water for a few days after the first of the summer rains fell each year. A week later, Rob was at the provincial health department launch of a water-awareness campaign to educate rural people not to use river water for five days after the first rains of the summer season. The department did not offer rainwater puddles as an alternative and had no knowledge that this practice – which would avoid the collection of bacteria-laden river water – was an indigenous way of the past and was still evident in some rural areas.
- A researcher investigating "wild spinach" used by rural communities (Mtshali, 1994), came across the indigenous practice of always throwing down *(ilahla)* organic waste at a particular spot in each rural homestead. This organic composting practice created a garden of edible wild plants. The wild garden enabled women to know and to show younger women the wild vegetables that are available and should be picked at differing times. Despite this knowledge-laden indigenous practice, education programmes introducing deep-trench agriculture as a permaculture technique either did not know of or failed to

acknowledge these processes as earlier and indigenous forms of organic farming in southern Africa.

It is somewhat ironic that state and NGO experts developing programmes to facilitate better rural health should still be relatively blind to a capital of contextual knowing amongst their intended target groups.

A methodological sense of indigenous knowledge in environmental learning

The case of *Nagana* uncovered in the Colenso diaries and the more recent cases of common-sense practices which remained unknown in expert contexts suggest that much of the capital of indigenous knowing has been overlooked or overshadowed against the conventional wisdom of institutional knowledge in the present day. Fragments of past and present indigenous knowing reside in everyday contexts. These embody implicit and explicit knowing in context, much of which was earlier sequestrated[5] to, and subsequently differentiated as, abstracted knowledge by experts operating in professional contexts outside everyday life.

Evidence in documents and from community interactions all point to how an engaging re-appropriation[6] of indigenous knowledge might be possible through processes of open deliberation that critically interact with existing knowing in context and knowledge that has come to be held outside everyday life. Beck (1992) describes this mediating mix of social processes as an "inter-epistemological dialogue" – meaning-making struggles between common-sense ways of seeing things in everyday life and knowledge differentiated in institutional settings outside everyday life. In inter-epistemological dialogue at the interface between the relatively blind practices in and of the everyday and outside institutional knowledge, one often finds the necessary discontinuities and contested fragments of knowing useful for processes of knowledge creation (or learning). Bauman (1993:147) offers that: "It is the misunderstanding that requires explanation, makes us pause and think, sets minds moving, triggers the process of conscious knowledge-building".

Much of the common-sense wisdom that was opened up in the environmental learning activities we engaged in over the past years was constructed at such interfaces of tension. This suggests that useful methodologies should articulate processes of inter-epistemological dialogue between traditional or everyday – and scientific or institutional knowledge.

If much of the wisdom in indigenous ways of knowing has been, and can easily be, overlooked, how might the associated discontinuities and meaning-making challenges be mobilised to enhance environmental learning within the school curriculum? This question is not easily answered without further case evidence from which we derived methodological insight into the re-appropriation of indigenous knowledge as a rich capital for learning in the present day.

Case stories of the re-appropriation of indigenous knowledge

The first Share-Net resource on indigenous knowledge was a joint initiative involving Mba Manqele, her mother and Rob. The materials were developed around descriptive evidence of practices in everyday rural life. Mba collected stories from her mother and she and Rob discussed aspects of rural life, for example, water collection and the making of *amasi* (soured milk) and sorghum beer. The project developed as a resource development activity with the English Language Education Trust. The outcome was some easy-reading, second-language learning materials for use in adult literacy classes. The activity provided methodological insights that then shaped other materials and methods for teaching and learning as well as further resource development and research by Lynette Masuku van Damme (1999).

Water collection: probing common-sense wisdom in past practices

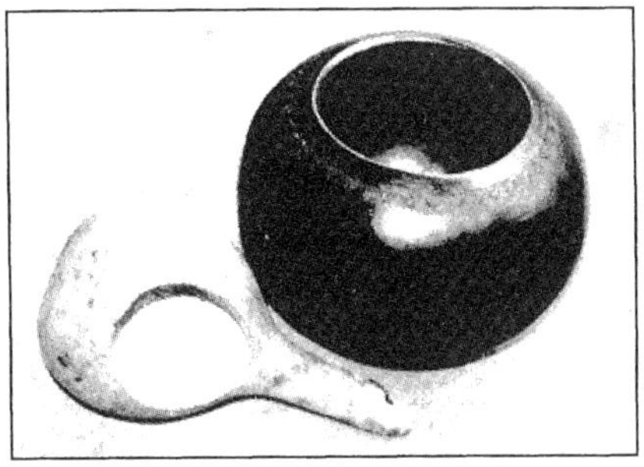

In the case of water collection, for example, Mba quite simply asked her mother to give her an account of rural water-collection practices from her youth. All of the details were recorded and ratified with Nguni colleagues to build a picture of "the how" and, where possible, "the why" of rural water collection practices.

Mba was often despondent as she struggled with fragments of evidence (what was done), shallow explanation and vague myth (why). These did not hang together or provide insights on the wisdom in and usefulness of many of the activities. The process was also often disappointing, not matching the hype around what was meant to be "IK". It was only as we began to build a picture of patterns within and behind common-sense practices that we were able to construct, ratify and narrate (re-appropriate) the common-sense steering knowledge embedded in many of the water collection practices of indigenous peoples of the region. By undertaking this interpretative work in close interaction with Mba's mother, it was possible for the three of us to reconstruct and to probe much of the inter-generational common-sense wisdom within the relatively blind habitual practices of many daily activities.

Some of the insights derived from our interpretations were:

- the circular sweeping of surface waters in the collection of sweet water created a "hole" within which bacterial loads are 20 to 30 per cent lower, a pattern ratified in bacterial counts during participatory research on water quality using low-cost water test kits;
- the use of clay pots that were slightly porous meant that the water cooled as it was carried home. Faced with dark, cooling conditions inside the pot, bacteria encase themselves in the pores of the pot and secrete a mucus that hardens into a calcareous sheath, further reducing bacterial numbers in drinking water. These

insights were provided by the scientific staff of Umgeni Water, a group with whom we cooperated on water quality work;

- great care was taken to scour out water pots, notably the silvery white calcium on the sides, before collecting the next pot of sweet water from the spring. This practice effectively ruptured the calcareous sheaths within which bacteria had encased themselves.

On their own, these practices are of minor significance, but together they tell a much bigger story of knowledge-laden indigenous practice.

This re-appropriating interpretation of practical wisdom in ways of old was recorded as two narratives that became the basis for Share-Net environmental learning support materials. One detailed the practices as told by Mba's mother and the other sought to illuminate some of the compelling common-sense wisdom in these practices. Great care had to be taken not to assume that the common-sense wisdom illuminated in everyday practices was indigenous knowledge in the sense that it was explicitly known in the community.

These early knowledge creation interactions taught us that:

- those within a particular cultural perspective can initially be reticent and even embarrassed, often seeing ways of old as something to be discarded and left behind;
- initial discourses in the community's mother tongue were often bland and did not provide insights until probed and reflected on within wider evidence.

These insights lead us to approach knowledge re-appropriation as inter-subjective processes[7] within which we worked as a team "going there, doing that and gathering evidence" until insights began to emerge. "Been there, done that and have the evidence to prove it" became an open-ended methodological focus within our continuing, somewhat *ad hoc* work on indigenous knowledge. This challenging perspective was particularly useful when Lynette Masuku van Damme chose to investigate grain storage.

Grain storage: the guided reconstruction of a grain pit

Lynette Masuku van Damme assembled a team of co-workers to construct a grain pit under the expert eye of an elderly colleague, Sikhakhane. The initiative was part of a project to document the capital of experience and knowledge of elderly conservation agency colleagues before they retired. Sikhakhane had built a grain pit in his youth and agreed to supervise Lynette[8] to build a grain pit as a working exhibit at the Umgeni Valley, an environment centre in Howick in the KwaZulu-Natal Midlands.

Methodological propositions developed through these processes included two realisations. First, in earlier times, methods of teaching and learning were simply a matter of applying skills learnt while working with others and under the guiding eye of an authority figure with experience and know-how. Second, the construction of the pit was a cooperative process that was re-enacted with widely held skills and without explanatory guidelines, but within a steering mythology with wide margins for practical variation in differing regions.

The re-appropriation of indigenous knowing presented as processes of creative re-enactment amidst steering story, relationally constructed within the challenges of the times, and with a mediating figure in authority. African indigenous knowing, far from being static and unchanging, has high levels of local variation in a land of high biological diversity. There is also evidence of an adaptive capacity to engage new ideas.

Some developing methodological insights

This evidence from and the experience of constructing a grain pit may have contributed to Masuku van Damme (1999) concluding that indigenous knowledge does not appear to be distinct from other equally useful ways of knowing. There was a developing sense that inter-generational appropriations of knowledge in contexts of engaging challenge are alive with misunderstandings and discontinuities within which insights develop to be brought forward and narrated as knowledge. Insights accompanying the re-appropriation of the common-sense knowing in indigenous ways of old also had reflexive and inter-epistemological dimensions.

Beck (1999:120), elaborating on unintended consequences born of patterns of unawareness[9] in modernity, describes how "[t]he claims of differing expert groups collide with each other as well as with the claims of ordinary knowledge and of the knowledge of social movements".

An unintended outcome of enlightenment and modernity has been disembedded knowledge held in institutions outside everyday life, and a corresponding fragmentation of situated knowing of the everyday. A heightened sense of risk has often fostered hysteria amidst dogmatism and continuing unawareness in social movements intent on, for example, the re-appropriation of indigenous knowledge. Latent tensions and discontinuities across institutional knowledge (e.g. formal science), knowing in the everyday ("indigenous knowledge") and knowledge developing in social imperatives amidst people of like mind engaged in cooperative activity (such as conservation groups), are all rich in discontinuities with meaning-making potential. Deliberation (engaging thought and talk about) and a detaching vantage point (a reflexive standing back) appear to be central to processes of knowledge creation necessary for the re-appropriation of indigenous knowledge in educative processes.

In support of this, Bauman (1993) suggests that knowledge arises amidst reciprocity and discontinuities in everyday life. The shared meaning-making struggle experienced in the re-appropriation of indigenous knowledge in the above cases brought forth explicit insights that might in earlier times have been more implicit in habit and taken-for-granted patterns of reciprocity found in many early indigenous practices. Bauman (1993:196) cautions that emotion-laden experiences and insight amidst rediscovered ways of old should not be seen as an empowering re-skilling. He points to the outcomes of processes such as these as often being a package deal within new patterns of technocratic authority, warning that there are limits to modern reflexivity. Beck (1999), elaborating on unawareness and unintended outcomes, also exposes some of the limits of reflexivity.

Our methodological probing of indigenous knowledge creation processes and the curriculum further involved cases of learning support materials use and the clarifying of appropriate methods of teaching and learning to mobilise indigenous knowledge.

Case stories of developing curriculum materials and methods

In the social politics of post-apartheid transformation, there has been a strong call for the recognition and bringing forth of an African symbolic capital in the multicultural melting pot of schooling. There were also imperatives for a break with the educational tyrannies of the past, notably a curriculum based on rote learning and examination failure among rural communities in particular.

Within this milieu, the early indigenous knowledge materials distributed through Share-Net found immediate use in methods aiming to bridge the symbolic capital of the marginalised indigenous peoples and the conceptual capital of academic disciplines in South Africa's school curriculum. As traditional school subject disciplines gave way to learning areas with the advent of outcomes-based education (see Lotz-Sisitka, this volume), further methodological space was opened up for indigenous knowledge. Methods and methodological perspectives today, however, still remain undifferentiated and contested amid the mix of constructivism and intended outcomes that makes up Curriculum 2005 (Lotz-Sisitka, this volume). In the next section of this chapter, we draw out insights into methodology that support knowledge creation (learning) by examining methods and materials as these have come to be used in the mobilising of indigenous knowledge within environmental learning activities in current school curriculum settings.

Early Share-Net booklets with an indigenous knowledge focus were, as described earlier, dual text materials for adult second-language literacy training. They had the methodological character of giving away the tools for probing a wealth of wisdom in the cultural fabric of everyday activities. The booklets were extremely popular and were immediately used in wider teaching and learning activities in the school curriculum.

The seemingly uncritical adoption and use of the booklets in, for example, science and language curriculum settings worried us. The texts were interpretative appropriations of Nguni indigenous knowing that were not intended for presentation. The idea of presenting indigenous knowledge in interpretative texts was contrary to our developing sense of indigenous knowledge processes in educational settings, namely an inter-epistemological bringing forth and probing of implicit and explicit knowing apparent in ways of old. What had been laid out in description and interpretative text was now being taken forth and presented in ways that we thought might cut off meaning-making struggle as indigenous knowledge, especially when simply narrated to learners in what was a second language for most.

To find that quite the opposite was happening came as a bit of a shock to us. Exposure to the texts seemed to enable learners to mobilise and narrate other and wider aspects of common-sense practices, past and present. What appears to have helped these learning activities is how meaning-making was made explicit amidst the interplay of story and interpretation in the text. This "method" seemingly "gave away interpretative tools" for probing aspects of other common-sense wisdom in indigenous and everyday practices.

The remainder of the chapter examines further case experiences to illustrate how some materials and methods are playing out within the open-ended methodological terrain that has been laid out in the chapter thus far. We now examine cases of Share-Net materials and methods in outcomes-based education curriculum settings in schools.

It would seem that one of the more obvious curriculum spaces for indigenous knowledge processes is the constructivist strategy of mobilising prior learning, the tuning in and bringing forth of what is known and meaningful to the learners, so that the curriculum provides relevant learning challenges that engage and build on existing knowledge.

Mobilising prior knowledge

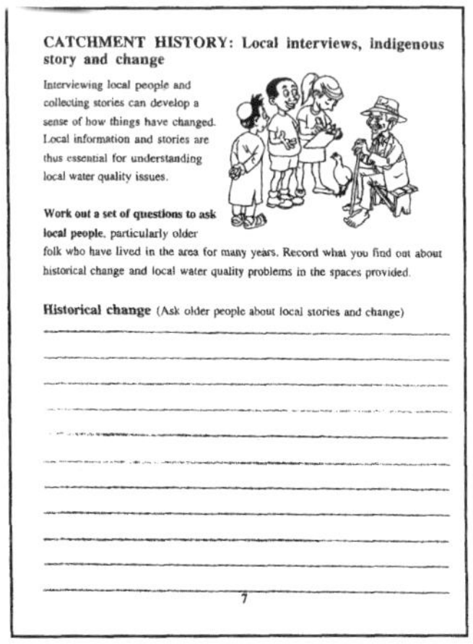
CATCHMENT HISTORY: Local interviews, indigenous story and change

Interviewing local people and collecting stories can develop a sense of how things have changed. Local information and stories are thus essential for understanding local water quality issues.

Work out a set of questions to ask local people, particularly older folk who have lived in the area for many years. Record what you find out about historical change and local water quality problems in the spaces provided.

Historical change (Ask older people about local stories and change)

7

Tuning in with a survey of story, practices and problems was used in a Water Audit resource pack for schools. The method was very useful for orientating learners in context but did not bring about much in the way of insights on indigenous knowledge.

The stories and information gathered by learners were superficial and often erroneous. The production of vague and undifferentiated ideas was in fact notable as the educative strength of the activity. Disagreement and the probing of ideas served to kick-start a learner engagement with the ambiguity and superficial pictures that had been brought forth by the investigation. Learners began to look wider and deeper, debating ideas and testing propositions, successively developing more reality-congruent pictures on matters in the local watershed. This suggests that educators should ensure a variety of probes for triangulation, to throw up inconsistencies and provide tools through which learners can investigate further.

The community and watershed survey was strengthened by the addition of a text on Nguni indigenous knowing and water collection. This became an information resource to support engaged meaning-making through the survey.

Providing information on patterns of indigenous knowing

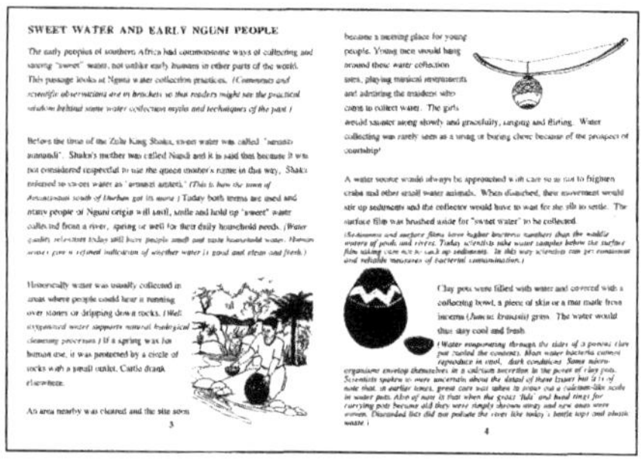
SWEET WATER AND EARLY NGUNI PEOPLE

Information-laden materials that illuminate knowledge-creation processes and challenge learners to uncover the wisdom in the everyday became the basis of useful methods. Used before the survey of local and past practices, these materials seem to provide conceptual scaffolding for documenting and probing indigenous knowing imbedded in past and current practices. The methods used to introduce the information have become many and varied. Some teachers give the materials out as a comprehension exercise, whereas others use story-telling approaches, particularly with young learners. One teacher Rob spoke with tells the story of practices, leaving out any interpretative insights, and then tells the story again. We have now taken this one step further and use pots and grass and act out the activities with these artifacts in preparation for fieldwork in local communities.

As more curriculum materials were developed and methods became many and varied, more diverse enquiry activities began to appear in the schools with which we worked.

Investigations drawing on indigenous practices of old

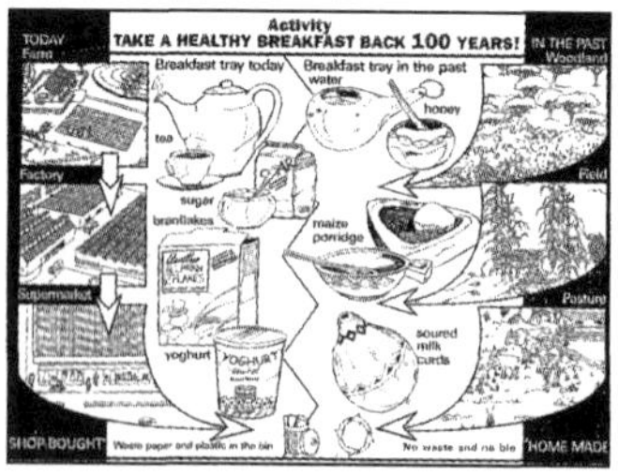

Enquiry (finding out) methods began to draw on indigenous perspectives and to set these against patterns of practice in the present day. The challenge of taking a healthy breakfast back 100 years was developed as a picture comparing life today with life in the past. Its curriculum use has been varied and rich. Notable has been how knowledge-creation processes have become more futures-orientated around the question: how can we blend present and past patters of life to come up with more sustainable ways of living together in our environment?

A notable limitation is that whereas all of the artefacts of earlier times were available to us, these are now only to be found in rural areas. Grindstone have provided us with some amusement. These are now used as bird baths by affluent urban people. Some of these have now been recovered to their intended use, but in curriculum settings.[10]

As knowledge-laden indigenous artefacts were sourced and used and compared with the ways we do things today, further action-centred materials and methods developed.

Using indigenous technologies

Actions that can be used to probe the wisdom in ways of old came to be introduced in differing curriculum settings. One of the most interesting examples of methods employing indigenous technologies is the *amasi* gourd (a calabash used for making soured milk). The spongy interior holds the bacteria so that the warm milk direct from the cow is kept warm and soured. A hole in the bottom drains off whey, which is retained for other uses. The indigenous soured-milk technology is a masterful design of elegant simplicity and ease of use with low impact on the environment.

Materials and methods in a curriculum framework

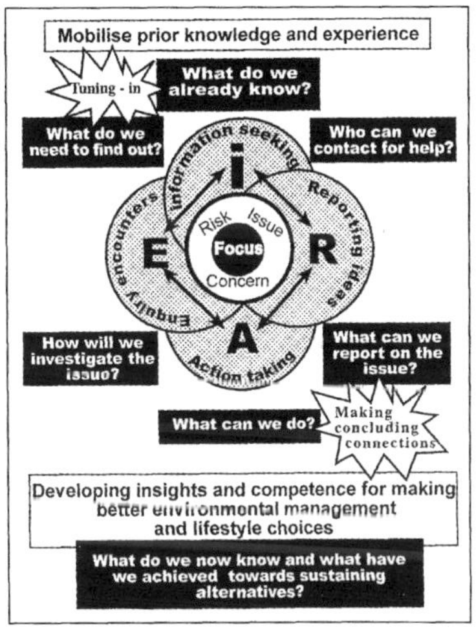

This latter section on materials and methods has been reported with reference to the dimensions of an Active Learning Framework: tuning-in, information-seeking, enquiry, action, reporting and making concluding connections. This curriculum planning tool was developed within the National Environmental Education Programme (O'Donoghue, 2001; see also Lotz-Sisitka, this volume). An active learning heuristic for facilitating environmental learning in outcomes-based education, it allows for a bringing-forth of indigenous knowledge in each of the intermeshed processes. It has been interesting to note how materials and methods have transgressed an initial expectation for indigenous knowledge processes to be located at the interface between everyday life-world experience and the curriculum and simply centred on a cultural capital of prior learning. We now see how an indigenous symbolic capital can have a much wider methodological scope in curriculum settings.

Some reflections on developing methods and methodologies

Environmental learning is proving to be a rich and challenging arena for mobilising and clarifying indigenous knowledge processes. The emerging evidence in this review suggests that indigenous knowledge is neither an objective property nor static. The developing picture also questions the assumptions that indigenous knowledge resides among and is the sole property of the indigenous, with knowing articulated in mother tongue and located in everyday cultural life.

In our examination of historical evidence and the probing of everyday life practices articulated in mother tongue, we have found that the terrains that provided a capital of indigenous knowledge were cross-cutting interactions and inter-epistemological challenges. These have arisen amid evidence of knowing in the everyday, expert institutional knowledge and the environmental movement within which contested and contrasting stories have come to interact and be interpreted.

Contrary to colonising sequestration of the past, what we have attempted is to give away interpretative tools and to locate the emerging narratives in open-ended, interpretative re-appropriation. The trajectory of these narratives is to make reflexive meaning-making processes freely accessible within a hoped-for renaissance of African discourses that might map out and realise more just and sustainable futures for all our peoples.

Endnotes

1 Share-Net is a publishing initiative convened by The Wildlife and Environment Society of South Africa for the production of low-cost learning support materials for environmental education.

2 This paper is a substantially reworked version of the paper presented at the University of Pretoria seminar on Endogenous Knowledge and published in Crossman, 2002.

3 Later reviewed and published in an EEASA monograph, O'Donoghue & Janse van Rensburg, 1999.

4 For an expanded review of the *Nagana* and diverse social processes that shaped the marginalisation and appropriation of indigenous knowledge into institutional settings in southern Africa, see O'Donoghue, 2002.

5 A term used by Bauman (1993) to illuminate institutionalising ruptures in the fabric of community knowing.

6 A term used by Giddens (1991) that encapsulates a sense of the power and property attachments associated with getting back that which was stolen from or lost to communities.

7 Meaning-making processes where ways of knowing are mediated in everyday interaction.

8 Having a woman leading and working on the project team was a radical departure from Zulu tradition, within which it was taboo to have a woman involved with anything to do with Nguni cattle.

9 Simply put, things that happen beyond what we see and think is going on.

10 A lateral-thinking learner has even suggested that the birds may have used them more than indigenous peoples did in the past, given that grind stones were used out in the environment and not often carried back to be located at homesteads until the wider use of grains like maize. It was simply more convenient to have the grindstone at the food source for ease of harvesting and grinding.

CHAPTER ELEVEN

Sustainable Development in a Post-Colonial Context:
The Potential for Emancipatory Research

Tsepo Mokuku, Lesotho

Introduction

This chapter explores the potential of "post-colonial action research" to facilitate a critical and emancipatory education which will promote sustainable development in Lesotho. It is argued that the post-colonial social order is characterised by social problems and contradictions, and that education, as a socio-cultural development process involving the transmission of predetermined knowledge systems, contributed to de-development and disempowerment. Focusing on selected cases of participatory action research (PAR) studies in Lesotho, it is further argued that whereas PAR has had localised benefits, it has failed to engage structures and to facilitate emancipation when there had been the opportunity to do so.

Lesotho's post-colonial context

The socio-political reality in Lesotho is characterised by many ambiguities. It is fussy, complex and difficult to analyse. Historically, Lesotho shifted from a period of prosperity, based on Basotho's high level of agricultural skills and knowledge, from the early 1800s to the early 1900s (Eldredge, 1993; Ferguson, 1997; Murray, 1981), to being classified as a "least-developed" country. It could therefore be described as having undergone a process of de-development over the last century, which has been associated with colonialism and industrialisation (Keegan, 1986), environmental degradation (Eldredge, 1993) and, more recently, with development programmes (Ferguson, 1997; Quinlan, 1995). Some of the other contradictions are the dominant education system that has benefited many and marginalised many more from the economic mainstream; many have gained from the dominant Christian religion for their socio-cultural development and many were left disillusioned; some groups have benefited from the post-colonial governance whereas many others have drifted to deep levels of deprivation; reliance on externally developed products and knowledge systems encapsulated in a colonial language has benefited the majority, but has also given rise to erosion and subjugation of local knowledge systems and language. These conditions constitute part of our daily experience and shape our consciousness and world view. They are so familiar that they prevail largely unnoticed.

My engagement in this context with action research and associated socially critical literature brought these contradictions into sharp relief for me, developing a great discomfort in my consciousness that in part motivates my attempt here to theorise a "post-colonial action research" that can promote sustainable development and emancipation in our country. Contradictions and ambiguities may proffer opportunities for new understandings of what development could mean in a post-colonial context, as well as opportunities for the exploration of new practices for development.

Lesotho's socio-cultural "development" initiatives in the realms of formal education, community development projects and religion have yielded in my view more complexities than clear-cut benefits and solutions. Cognisant of the complex of forces that shape educational development initiatives, I argue that this has primarily been due to trends in education that are underpinned by enlightenment assumptions of education: typically some predetermined forms of knowledge are imparted by the "enlightened" to the "ignorant" or by the "educated" to the "uneducated". In their endeavour to enlighten, colonists and missionaries delivered "packages" of knowledge to civilise the barbarians. In the post-colonial context, the urban educated and enlightened deliver development programmes to the uneducated rural communities in order to develop them. In formal educational settings, learners are imparted with curricular knowledge designed to educate and enlighten. Thus education in this context has mainly assumed an impositional, imperialistic and colonising function characterised by uncritical transmission and assimilation of predominantly European and American knowledge systems and models. The now dominant super-structures or paradigms of knowledge transmission, the education systems and the Christian religious institutions, emerged with the arrival of the missionaries in 1833, to be fortified during almost a century of British colonial rule (1868–1966). A school and a church were aligned strategically to educate and enculturate, laying a substructure for an efficient knowledge delivery system that is very active to date. These structures of education have also historically assumed a form of post-colonial or North-South "pump" that actively transfers European knowledge to Lesotho. Typically, the educators' function in this system is to transmit predetermined knowledge forms, developed outside Lesotho, to the learners and thus shape their world views, usually in ways that are incongruent with local realities, giving rise to a vicious circle of dependency on externally developed knowledge systems, which in turn shape local social reality. This process characterises diverse learning environments in Lesotho: schools, universities, seminaries, churches, workshops, short-term training programmes, community development programmes. It has arguably empowered a few, and largely:

- psychologically disempowered many people and inhibited their creativity to respond to their problems and challenges;
- subjugated local knowledge/thought forms and developed an inferiority complex in the majority of the Basotho about their local knowledge (embodied in their language and culture);

- advanced European-derived knowledge systems, to the detriment of local knowledge systems;
- developed an inferior, brittle and disoriented culture.

The transmission assumptions of education also seem to characterise development work in Lesotho. For example, some conservation and livestock development programmes were undertaken by the British colonial power in the early 1900s, primarily to control unemployment, rural-to-urban migration and the burgeoning migrant-labour system (Driver, 1998), and later by the aid organisations (Ferguson, 1997; Quinlan, 1995). These initiatives largely involved the application of technical solutions based on cause-effect analyses, without any regard to the intricate interactions of the local livestock owners with the environment and the knowledge they had acquired over years (Quinlan, 1995). Typical intervention strategies have included predetermined solutions such as transmission of messages to rural communities to de-stock (Quinlan, 1995), construction of contour banks, and fencing in of denuded range land (Ferguson, 1997; Driver, 1998). In an interview, in 1951, on the apparent (short-lived) success of the contour bank implementation scheme, a colonial Agricultural Officer stated: "We explain the method to the chiefs, and the chiefs pass down our orders to the villages and see that the orders are carried out" (Driver, 1998:60).

Post-colonial action research

The origin of action research may be traced to the work of Kurt Lewin, in 1946, and other Westerners, undertaken as a problem-solving method to emancipate deprived communities from their social problems (McKernan, 1991; Winter, 1989). Participatory action research (PAR) has strong roots in the African context. Early PAR initiatives in Tanzania benefited from the work of South American education activist Paulo Freire and North American development researcher Budd Hall. The earliest recorded application of action research in Lesotho is the work of Janet Stuart in the early 1980s. My first encounter with the method was in educational institutions in South Africa in the early 1990s.

Participatory action research is purported to raise the researcher's "consciousness" and to enable "emancipation" from oppressive conditions (Grundy, 1987; Tripp, 1990; Walker, 1990). Carr & Kemmis (1986) have distinguished between technical, practical and emancipatory action research, and Zuber-Skerritt (1996:4–5) has summarised these as follows:

> *Technical* action research aims to improve effectiveness of educational or managerial practice. The practitioners are co-opted and depend greatly on the researcher as a facilitator. *Practical* action research, in addition to effectiveness, aims at the practitioner's understanding and professional development. The researcher's role is Socratic: to encourage practical deliberation and self-reflection on the part of the practitioners. Action research is *emancipatory* when it aims not only at technical and practical improvement and the participants' better understanding, along with transformation and change

> within the existing boundaries and conditions, but also at changing the system itself or those conditions which impede desired improvement in the system/organisation ...

PAR studies in Lesotho to date may be described as technical and practical but fall short of taking up the challenge of emancipation. Well-documented projects in this area include curriculum development-related studies at primary, post-primary and tertiary educational institutions; a large-scale action research project undertaken by the Institute of Education with primary school teachers in Lesotho; a project I undertook with a team of secondary school science teachers; and a number of small-scale studies undertaken annually by Bachelor of Education students in the Faculty of Education of the National University of Lesotho. The latter will not be the focus of this paper. I critique the three major educational action research projects in Lesotho and theorise on a post-colonial action research – action research that could be emancipatory, socially transformative, and in the interests of contextual sustainable development.

The Lesotho Action Research Group Project (LARGP), undertaken by Stuart, Morojele and Lefoka between 1984 and 1985, is the earliest documented project of its kind (Stuart, Morojele & Lefoka, 1997). It focused on practical and professional issues in the classroom, and involved teachers employing the method to improve the teaching and learning of a newly introduced subject, Development Studies. The epistemological base of the subject allowed for critical inquiry which fitted well with action research (Stuart, Morojele & Lefoka, 1997:166). Learners were encouraged to shift from a fixed and closed view of knowledge to an open and emergent one. Through recurring cycles of action and reflection they came out of their passive, receptive and silent mode to ask more questions, argue and discuss. Over time, they were able to show evidence of the ability to respond to higher-order cognitive questions. Participating teachers began to theorise about their own teaching and extended their repertoire of pedagogic methods. Through action and reflection, they shifted their emphasis from remedying what they saw as deficits in students towards analysing their own teaching.

However, the inquiry did not transcend disciplinary content so as to engage the socio-political context and structures in which the teachers operated. For instance, in a lesson where learners were given a task to discuss the topic "What would happen if South African mines were closed down and Basotho men sent home?", learners claimed that they "could not express themselves fluently in English". The study did not go beyond simply noting that English was a problem (Stuart, Morojele & Lefoka, 1997:175,177,185). A post-colonial action research process could interrogate structures that legitimise the use of English as a supposedly empowering language when it actually inhibits learning. Struck by this contradiction in the post-colonial nation states, Ngugi wa Thiong'o has observed that knowledge in these environments is "caged within the linguistic fence of its colonial inheritance" (1993:22). Emancipatory action research – particularly in a Development Studies classroom – could engage students to explore the origins of the English language in Lesotho and make apparent the colonial linkages of the language with their struggle to learn. This realisation would awaken learners to the legacy of colonialism and how it impinges on socio-cultural development. Based on classroom praxis and evidence of improved performance with the use of a mother-tongue and bilingual studies in this context (e.g. Nthathakane, 2001), the class could interrogate the assumptions that English is an appropriate medium of instruction. They could develop their own language policy, and thus begin to make a contribution to potentially transforming the British-modelled education system.

The development of higher-order cognitive skills and the ability to argue, which Stuart and colleagues initiated in their research, could be explored praxiologically in the interest of empowering learners to develop and defend aspects of indigenous knowledge systems. Recognising how Basotho were unable to defend their knowledge systems and culture on encounter with the colonists, higher-order reasoning becomes an imperative, not for answering challenging examination questions, but for emancipation.

Eugene Casalis, one of the three young French missionaries who arrived in Lesotho in 1833 as part of the expansionist Paris Evangelical Mission Society, provides an informative narrative which portrays Basotho as passive and amenable to cultural imperialism. With an open-minded perspective to Casalis's religious ideas, Moshoeshoe[1] had invited him to perform a religious service at his wife's burial. As custom dictated, a grave had been dug in the cattle kraal, and a number of cattle were ready for slaughter as an offering to the ancestors. For Casalis, such an amalgamation of systems was unacceptable. Moshoeshoe's position had to be "black or white". A debate ensued and Basotho lost.

Casalis: ...These cattle prove that you are tempted to pray to my God and to your *barimos*[2] as well. I should be unfaithful to my Master if I acceded to your desires.

Moshoeshoe: I told you so (exclaimed Moshoeshoe turning to the crowd), it cannot be.

Crowd: [His words were received with a murmur of discontent; the brother of the deceased protested against deviation from the national customs.]

Moshoeshoe: On what are these customs founded? *I should like to see the book where they are enjoined: the missionaries give us the reason* for all they do....

Ratsiu [the chief opponent]: *We are silent because we will not yield.*

Participant 1: What the missionaries say would be excellent if we believed it; but I, for my part, do not believe it.

Moshoeshoe: Nevertheless it is the truth, Yes the truth! The truth!

Casalis: Great men of Thaba-bosiu, aged men whom we all respect ... *[r]efute the truth that I declare to you*: here I am, speak: I listen.

Moshoeshoe: And I also listen, speak.

Crowd: [A long silence].

Participant 2: *We will speak when the missionary is gone.*

Moshoeshoe: Yes, you will conquer when there is no longer an adversary; *speak now!*

> *Why do you hang your heads?* I said in my heart, there are round me men who have words of wisdom ... let this pit be filled immediately, and the cattle be driven to the fields ... (Then turning to me [Casalis] he said) You have conquered ... (Casalis, 1861:90–91, my emphasis).

The silence of the crowd and their inability to debate and defend their knowledge suggests that they were being drawn into higher-order reasoning about their religion, which was apparently not underpinned by rationality, the lack of which rendered the religion amenable to refutation. A view of knowledge as fixed creates conditions for a closed, silent, or passive culture – perhaps the culture that Casalis encountered.

From the post-colonial perspective, this strategy of religious education for the socio-cultural development of Basotho set in motion the demise of Basotho religious knowledge and their reliance on an externally developed system of religion. A contextual participatory process of socio-cultural development, however, where communities engage in reflective processes of action and reflection to develop and refine the understandings and meanings underpinning their religious practices, would have constituted an authentic and sustainable socio-cultural development of this nation. Our engagement in this process today would constitute what Kincheloe (1995:75) refers to as a "sociocognitive emancipation", a first step towards "cognitive revolution".

The Primary Teachers' Project (PTP), facilitated by the Institute of Education of the National University of Lesotho, is another example of a large-scale PAR project undertaken at primary schools in Lesotho, in 1990 (see Stuart, Morojele & Lefoka, 1997). The project received government support, and had government-employed district resource teachers and supervisors participating in the capacity of facilitators. This project also focused on the professional development of teachers and was intended essentially to develop their capacity to reflect on their practice. Over 100 teachers were able to produce reports of their practice, rather more narratives than systematic research reports based on data collection and analysis, which were edited and published. However, this study too was neither socially aware nor critical, and did not reach beyond the confines of the school context.

Between 1996 and 1999, another PAR project entitled *Towards Education for Environmental Literacy* was undertaken and documented by me in collaboration with a team of three science teachers from three different schools, the Masianokeng Environmental Centre, individual staff from the National Curriculum Development Centre, and staff from the National University of Lesotho (Mokuku, 2001). Whereas this project was also classroom-based and focused on methods in the teaching of science, it also provided useful insights into possible directions for emancipatory post-colonial action research. At the beginning of the project we planned to undertake recurring action research cycles with two objectives in mind:

- to develop the meaning of environmental literacy in the context of Lesotho;
- to explore how science may be taught in accordance with the emerging meaning of environmental literacy.

PAR and the development of environmental literacy shared common goals:

- empowerment of research team members and learners for self-determination;
- social change;
- emancipation from dependency and externally-driven knowledge and theories;
- a holistic view of social reality.

The emancipatory and socially transformative perspective, enthusiastically embraced by the team at the beginning of the project, was unfortunately lost. The assumption that it was in fact possible to develop – in the English language – a theory of the concept of environmental literacy (developed in the USA by Roth, 1992) that would be responsive to the local environment proved to be mistaken.

The situation approximated what McTaggart (1997) called "academic imperialism". It is therefore not surprising that the emerging meaning of environmental literacy consisted of propositions of words associated with environmental literacy, rather than of any coherent theory. In the participants' struggle to verbalise their conceptions of environmental literacy in a second language, and in their drawing on their teaching experience, environmental literacy became associated with "taking action" to solve environmental problems, where action was understood to include a "discussion" of environmental problems, and the concept of "voluntary action" was understood as the opposite of compelling students to collect litter on school grounds (Mokuku, 2001).

The project drifted into an uncritical, technical mode. The teachers reflected on their class experiences to reconstruct their understanding of environmental literacy. They drew on outcomes of new teaching approaches they had tried, but did not address core structural and contextual issues such as the use of English as a medium of instruction when students discussed environmental issues, the shortage of laboratory equipment, and tightly structured teaching programmes which stifled creative teaching initiatives.

A second important assumption that lost the socially critical dimension of the research process was the introduction of PAR to teachers as a step-wise method of recurring cycles of planning, action and monitoring and a subsequent rigid adherence to the "four moments" of action research. One participant came away feeling that PAR was nothing more than a self-evaluation procedure. Another assumed that recurring cycles meant that she had to repeat the same content in teaching subsequent lessons.

A dream that I had in my wrestle with the concept of action research during the study provides a useful narrative for introducing emancipatory action research in a post-colonial action context (Mokuku, 2001):

> I was on top of a very high mountain with some people, holding a big book in my hand. This book was my power and my strength and I could not part with it. Suddenly the book fell out of my hands and I helplessly watched it tumble down the mountain slope. I saw it get smaller and smaller, until it became a tiny, shiny rectangular spot at the foothill. Just when I thought I had my eyes fixed on it, I saw many other similar spots;

> they were roof tops of houses scattered at the foothill. I hastily started to find my way down the mountain and ran at full speed to retrieve it. This turned into a long journey through a dense and scary forest; when I got to scary places I ran faster. As I took a turn at the end of a slightly steep path, I suddenly met with a black man with grey hair sitting down by the side of the path. He was friendly and appeared to be filled with wisdom and even to know everything about me. He used some English words when speaking to me. As we were talking, he assured me that I was not lost, and that I had not lost any treasure: "stop running!" he said. "Don't you know that your journey is the knowledge and the wisdom you are looking for?"

Based on this narrative, PAR may be introduced in this context as a process of moving away from dependency on pre-determined knowledge forms to explore the unknown for an unexpected wisdom to emerge from a "slow" or consciously reflective journey through the social surroundings. It is not a mapped-out journey of fixed steps to search and locate any treasure known to be lost. Research is not a set of procedures or a step-wise methodology to be implemented. If conceived as the latter, it becomes impositional and loses its emancipatory potential.

I have found an "open-minded" and "open-eyed" approach to what counts as evidence in action research[3] to be an exciting aspect of PAR that has enabled me to draw on my feelings, impressions and dreams. From this data may be constructed what I call "pono-tebo" or "pono-validity", a concept that could guide deep engagement with invisible forces of post-colonial structures. It is the researcher's *sensitivity* to the interplay of non-empirical experiences deriving from intimate interaction with context, recognising the associated emerging insights or visions illuminating instances of the researcher's concern, as if glimpsing beyond the empirical world that frustrates the grasp of the actual reality. From the sharp insights or visions a narrative that provides rich material for a critical reflection on aspects of the research may be developed – I have found my dreams a powerful material in this regard. Pono-validity is essentially one of the indicators of the researcher's deep attachment to the context of study, that deviates from positivist criteria for validity where detachment from the research context is valued for objectivity. Pono-validity does not devalue empirical validity; rather, it complements it. The attachment to the process of research and context presumes that those who experience and feel subjugation can better emancipate and develop themselves.

> Another dream:
>
> I saw myself with many people. We were at a place of learning, and we were being taught. When we were taught I developed a feeling of dissatisfaction with what we were being taught, and then I decided to part with that group and that place.
>
> Then I gave myself to the land. I covered the plateaus. I was coming from the west going to the east. I remember feeling scared; I was alone in a stretching land; my speed was like

> that of Nts'oekhe. I found myself overlooking the mountains, with huge cliffs, and deep scary gorges. I took courage and went down. I was painfully scared, but I just went on, even though I did not exactly know where I was heading. I ran on and on and then I saw a village; it was at a distance down there. When I came closer, I found that the village was inhabited by Africans. It was built of stones, and thatched with grass. The people there were satisfied, were not in a hurry. Peace reigned. I remember seeing a wide path, more like a road, but there was no traffic of cars. I was impressed by the satisfaction of the people of that village.

Based on my dream narrative, I developed a conviction that the attainment of an appropriate education system in this context involves the courage to shift from the established socio-cultural order and to explore hidden, alternative ways of living that place the interests and culture of people at the centre of education and development. This points me towards a direction for action research that serves emancipatory interests by encouraging inquiry to move "towards an understanding of the social and cultural structures that help shape the educational lives of individuals and help determine the consequences of schooling" (Kincheloe, 1995:82). This contrasts with the dominant colonially rooted system of education wherein research reproduces the *status quo*:

> No African living in the 20th century has escaped the taint of the colonial experience. None has freed himself from the colonial mentality in every department, or from the structures which maintain and reproduce that mentality. The decolonisation of the African mind must therefore be seen as a collective enterprise, as a communal exorcism through an intellectual bath in which we need one another's help to scrub those nooks of our minds which we cannot scour by ourselves (Chinweizu, 1987:9).

It would seem that my employment of PAR took me through the experience described by Kincheloe concerning critical action researchers: "... they awaken from the modernist dream with its unexamined landscape of knowledge, and unimaginative consciousness construction. Once awakened, critical teachers as researchers begin to see schools as human creations with meanings of possibilities lurking beneath the surface appearances..." (1995:77). It was perhaps because of this effect that I found the ideas of Machobane[4] relevant for the study, and invited him to our research team meeting. This is what he said:

> We are a country, we are a government. A country, which is us, we are fed by other nations through aid programmes. This person who is helping us is coming! And he is just around. The first thing he is going to take from us is women. Already they are crossing the border. Yet a country without women is finished! And the second thing is men. There they are, already in South Africa. And the men who are left here at home, you will see a man climbing a hill going up there; he is hungry! Now when he is there, a parliament sits [starts talking to himself]. When he comes down, he sees a helpless

> woman there at a distance. He looks at her, he keeps looking at her; after sometime they arrive, and surround her there [i.e. rape her], as if they were dogs fighting over a bone. These are the things that are happening right now, they govern us in Africa, the entire continent.
>
> This slavery started in the 14th century. This Africa, north Africa, had cities, big cities. Those people governed themselves. Then the white people came gradually, and then we started trading with them. Then arrived the Arabs. They came with sugar, in their ship. Then started a trade of black slaves to America. Then we were sold, the entire continent! Once this was stopped, came what is called colonialism. They are still doing it today. Then they developed a structure which is "curriculum" so that they govern us and make us their servants. When they had developed curriculum for us, then we fastened ties. We buy clothes from them, and we must never invent anything! Basotho used to smelt ore, smelted spears, doing all sorts of things. They invented. Then there was an abolishment of his creativity, so that he may be employed ... so that when they beg [for a job] they may say or write, "Vacancy!" And then write, "I beg to apply for a vacancy". Then when he ends he says, "your most obedient servant". What I mean is that our entire structure has been taken away from us.
>
> This is why I decided to take the kids, from a young age, to learn how to milk, together with his/her mother. I taught even mathuela, and said a person practices witchcraft only because he is hungry. It is hunger which underlies all things, and this is only because the curriculum was not developed by the people themselves. Let us sit down to make curriculum. That is why I established a college [Mants'a tlala College]. Today, a person who has been through secondary school, a girl, is a maid, a boy steals because he can not work anywhere. A matric graduate dislikes milking, he does not want to do anything with his hands, he is expecting to eat this beautiful milk of Longlife5 which has been made by others on that end. Then the Afrikaners took this land. When they had taken the land, they ploughed. These Basotho then continued to be taught in schools, to learn how to sing hymns. They know hymns so well. Then the Afrikaners continued to farm on that end. They are still in control of farms. They are the ones who developed the economy [in South Africa].

This critical historical narrative had a conscientising edge, which led one of the participants to remark: "everything that is given to us comes from somewhere else and ... all we need to do is just follow". His narrative had illuminated the historical context of the education system, shedding light on a world that normally remains hidden from educators. Kincheloe argues that "whenever we dispense with values, political considerations or historical context, our attempt to understand the situation we are researching is weakened. Our appreciation of an educational situation is contingent on the context within which we encountered it ..." (1995:75).

Conclusion

An emancipatory post-colonial action research orientation emerges here as an alternative to educational processes for development that are impositional and imperialiastic. It is a critical approach to action research wherein groups of people collaboratively engage in a contextual, socially critical process of action and reflection while addressing issues of their concern. It does not stop at exposing contemporary structural-ideological forces that shape the present reality/context and the consciousness of those engaged in the research, but moves on to uncover the complex of their historical roots and to reflexively and courageously transform them. To achieve this, it draws on critical historical narratives that make apparent the relationship between knowledge, power and domination. The telling of stories of those who are marginalised, and the critical historical accounts of how things came to be, is the beginning of the journey of emancipation, a first step towards sustainable development. This process of critical reflection on and reconstruction of context constitutes an important ingredient for sustainable development. Action research processes can contribute significantly to such ends, but a rigid technical application of PAR as a step-wise procedure would, to draw on Ngugi wa Thiongo's (1993; 1996) logic, render the process "imperialistic" and thus "trap" it from initiating significant socio-political transformation.

Endnotes

1 Moshoeshoe was the King of Basotho. He had a high regard for Christian religion, although he did not accept it as a replacement of his culture and religion.

2 The word *barimos* should be *balimos* and means ancestors.

3 See McTaggart, 1997:37; Kemmis and McTaggart, 1988:13.

4 J.J. Machobane (b. 1914) is a novelist and self-taught agriculturist. In the 1940s and 1950s, shocked by the hunger in his country, he perfected a small-scale permaculture system for Lesotho conditions, and then established a college to teach his methods. The post-independence government of the 1960s was afraid of his popularity and suppressed his methods, forcing Machobane to go into hiding for many years. In the 1990s, his system was revived, and is now practised by over 2 000 farmers in Lesotho, with NGO support.

5 Longlife milk is a commercial product and one of the many foodstuffs exported to Lesotho. It is specially packed to have a long shelf-life, without any need for refrigeration. This and other import products are associated with modernism and the capitalist economy which emerged rapidly in Lesotho from the 1870s onwards (see Keegan, 1986).

CHAPTER TWELVE

Ambivalent Globalising Influences in a Local Context:

The Case of an Environmental Education Practitioner's Experience in Zambia

Justin Lupele, Zambia

> What is required is a "politics of the rhizome" in which everyone, every community, every nation and every major region comes to terms with the uniqueness of their situation and acts accordingly – and realises that it is only by vast numbers of people, of groups and organisations acting in accordance with the possibilities of their situations, that the environmental crisis will be solved. The actions required ... will differ according to their situation within the global economy ... it is necessary to identify forces at work affecting individuals ... nations ... and regions ... a knowledge of the dynamics of the global system ... is required (Gare, 1995:161).

Introduction

The past 30 years have seen increased global attention being given to environmental issues and risks. Their spread has occurred alongside the expansionist nature of the information revolution and much of this has reached African countries through neo-colonial and post-independence development aid initiatives. I work as an environmental education practitioner in Zambia for the World Wide Fund for Nature (WWF), an international conservation organisation which has operated in Zambia for over 40 years. This paper highlights some features of the changing face of conservation in Africa. It also shares some insights on the ambivalence of global influences on the local context, in particular, how global influences have both positive and negative, as well as unintended effects.

In particular, I wish to highlight the following:

- the influence of global policy developments on national and local policy-making;
- the influence of development aid on possibilities for practice;
- the influence of educational concepts and ideas on local practice;
- local voices and perspectives influencing ways in which Community-based Natural Resource Management (CBNRM) and education are practised;
- the influence of conservation ideals in the lives of communities.

Throughout the chapter, I indicate how these influences have impacted upon, or influenced, my practice, in the context of the WWF Zambia Education Project (WWF ZEP).

My work in the World Wide Fund Zambia Education Project

I am an education officer on the project, in charge of resource materials development in the formal and non-formal sectors. Over the past six years I have been involved in participatory materials development in school and community contexts. These materials are meant to enhance and contribute to the project's goal, which is "to increase capacities in rural communities in Zambia to address their environment and development of their natural resources for short- and long-term benefits aimed at alleviating poverty" (WWF ZEP, 2000:2).

WWF started conservation work in Zambia in 1962. This started with the purchase of the Lochinvar Ranch, from private hands, on behalf of the Zambian Government. The ranch was at the time teeming with a large presence of wildlife, especially the endemic Kafue Lechwe[1] and a variety of bird life.[2] WWF's earliest support in environmental education in Zambia was through the Wildlife and Environmental Conservation Society of Zambia (WECSZ) which, together with Bata Shoe Company, supported the launch of Chongololo[3] Club in 1973 (Carwardine, 1987). In 1980 a sister club to Chongololo, the Wildlife Conservation Club, was launched under WECSZ to cater for secondary schools and college students. A weekly 30-minute radio programme called Chongololo Club of the Air runs on Radio Two of the Zambia National Broadcasting Corporation, disseminating environmental awareness messages to a wide audience. WECSZ has been involved in the development of education materials (with the help of WWF and other well-wishers) in the form of books, posters, leaflets and pamphlets to raise awareness of various environmental issues. The Chongololo Club was one of the first means of providing some form of environmental education (Carwardine, 1987). For over a generation the Chongololo programme has been trying to put nature on the local and national agenda (Thompson, 1998).

In 1989, WWF International responded to the Zambian government's need for a comprehensive environmental education programme as identified by the National Conservation Strategy of 1985. This gave birth to WWF Zambia Environmental Education Programme, now called WWF Zambia Education Project (WWF ZEP). The project's initial activity was the production of materials for primary schools, to meet the dire need for educational materials.

Key among international developments in the field of environmental education is the realisation that environmental education processes are not simply about nature awareness. These changes have influenced environmental education practice in Zambia. For example, WWF ZEP has been working with communities in the area of environmental education, with the aim of assisting them to achieve the capacity to initiate and carry out activities that would improve their livelihoods and immediate environment. This involves engaging communities in learning processes that will empower and help them to identify available alternatives in the utilisation of natural resources. Through these educational processes, communities engage in actions that address their needs and contribute to poverty alleviation. Our experience in working with communities has shown that environmental issues as well as perceptions of sustainable development differ from one community to another (WWF ZEP, 2000). In terms of environmental education materials and training provision, each community group should be considered as unique, with its own needs.

Zambia

Zambia is a landlocked country located on the great plateau of central Africa at an average elevation of 1 200 metres above sea level. The highest elevations range from 1 525 to 1 650 metres above sea level, and are found in the Northern Province (Mbala, Mporokoso & Mpika), the Central Province (Serenje), and the North Western Province (between Solwezi and Mwinilunga). There is a montane zone in the northeastern part bordering with Malawi that exceeds 1 800 metres above sea level (Environmental Council of Zambia, 2001:2). The eastern and southern provinces fall in deep rift troughs, the Luangwa and Middle Zambezi valleys respectively, with altitude ranges of 350 to 600 metres above sea level (ECZ, 1994).

Zambia has 73 officially recognised ethnic tribes and seven official vernaculars that are used on national radio and television. The seven languages are also used as media of instruction in lower primary schools. Each of the nine provinces uses one or two of the vernaculars in primary schools. These are Chinyanja, Icibemba, Silozi, Chitonga, Kikaonde, Luvale and Lunda. Historical and archaeological research findings indicate that almost all the tribes of Zambia originally came from Luba Lunda empire in Zaïre, in the present Shaba Province (Machishi & Musona, 1999). They were called the Bantu-speaking people. These invaded and pushed the Bushmen out of their land. The languages spoken by many tribes in Zambia today are similar to those spoken by the Late Iron Age people, probably as a result of intermarriages between the two groups of people (Manda, Bwalya & Simbeya, 1997).

The country has remained a haven of peace in Africa for the past 38 years. The first republican president Dr Kenneth Kaunda dismantled the previously rigid tribal barriers through the philosophy of humanism. The country's strong national identity and pride is founded on Kaunda's motto: One Zambia One Nation (Carwardine, 1987). Inter-tribal marriages were previously unheard of, but are now commonplace. However, different tribal cultures still exist and the various tribes still practise their traditional dancing. Apart from the ethnic groups, Zambia has a small community of Europeans and Asians. The Europeans work in the mining and farming sectors while the Asians control trade and commerce. Zambia is also home to tens of thousands of refugees from Rwanda, Angola, Democratic Republic of Congo and Burundi. Most southern African countries owe their political independence to Zambia, which played a pivotal role in the liberation of the region.

Zambian agriculture is characterised by subsistence, as well as large and small-scale commercial farming. The rural dwellers are predominantly subsistence farmers who produce enough for their families and sell only the surplus. The white population dominates large commercial farming. Maize has always been the major cash crop, followed by sorghum. In recent years, export crops such as paprika have been grown on a large commercial scale. The country's staple food is maize, which is ground into maize meal (mealie meal). Mealie meal is used in the preparation of a thick porridge called *nsima*. Most households depend on *nsima*, which is eaten as lunch and supper with some relish such as vegetables or meat. The country is, however, not self-sufficient in food production because of poor agricultural policies and sometimes adverse climatic conditions.

Like in many southern African countries, Zambia is endowed with natural resources, which are communally owned with open access (Booth *et al.*, 1994). Miombo woodlands (a sub-category of the savannah) is the predominant vegetation type. This is part of the open forest. The others, in order of predominance, are grasslands, closed forest types and the termitaria bush group type. The natural vegetation cover intensity is high in the high-rainfall areas in the northern and north-western parts of the country. The intensity of the natural vegetation cover diminishes towards the low-rainfall zones in the south. The natural vegetation cover and the rich biodiversity are threatened by the socio-economic factors associated with the ever-growing population, which now stands at just over 10 million. The country is said to have one of the highest population growth rates (of 3.1 per cent per annum) in the world (UNDP, 1998). In economic terms, the country's biodiversity is a composite asset that provides a source of livelihood for the majority of the people, especially those in rural areas as they depend on the natural resources occurring in their areas for their livelihood (Hachileka, 1997).

WWF Zambia Education Project works in the Nalusanga community of Chieftainess Kabulwelwe of Mumbwa district, Central Province. The community members are involved in bee keeping, fish farming, and fruit-tree growing as alternatives to the exploitation of the natural resources. In the Kafue district in the area of Chieftainess Chiawa, Lusaka province, we are working with the Chiawa community in the running of a community campsite to help realise income potential from the wildlife resources in the area as well as develop appreciation of the importance of the animals. In Chief Chiundaponde's area in Mpika district, Northern Province, we have established an ecotourism project, beekeeping, rabbitry and fish farming, amongst the community members. In all these community activities, my role as an education officer in charge of resource materials development has been to facilitate the production of resource materials by the community members. These enhance the learning process and provide opportunities for discussions. The project is helping communities to increase their capacities to address their environmental concerns and to understand, initiate and participate in the sustainable utilisation and management of their natural resources for the benefit of conservation and to alleviate poverty (WWF ZEP, 2000).

Shaping politics and economics in Zambia

Global influences such as Perestroika, Mikhail Gorbachev's 1986 programme of economic, political and social restructuring, which contributed to the dismantling of the totalitarian state of the Soviet Union, sent a wind of political and economic change across the African continent in the early 1990s. More and more countries in Africa advocated for political and economic change to remove dictatorial and stifling governments that had long outlived their usefulness. Perestroika trickled down to us in Zambia, and in 1991 the country reverted to multi-party politics. Zambia gained her independence from Britain in 1964 and was governed as a multi-party state until 1972. From 1972 to 1991 Zambia had been a one-party state.

With Perestroika came the World Bank and the International Monetary Fund's (IMF) pressure on the bankrupt countries of Africa to adopt structural adjustment programmes (SAP). These programmes forced African countries to restructure their economies, reduce spending on health, education and other social services, in order to service their debt. The intention was to create an environment conducive to foreign investment and economic growth. The new political leaders in Zambia embraced SAP religiously. World Bank and IMF economists advised that the Zambian public sector was bloated and that companies would benefit from privatisation. Nearly all the parastatal companies were rapidly sold off. Consequently, the country experienced mass redundancies and unemployment. Over half a million people lost their jobs, and have been left virtually destitute. The number of street children roaming the streets of the major cities has increased dramatically. The situation was worsened by the drought that hit the country between 1991 and 1992, resulting in the impoverishment of many rural people, forcing them to dispose of important assets such as livestock and seed stock (Binns, 1995:126). The government was forced to import maize from other countries.

The *State of the Environment Report for Zambia* (ECZ, 2001) reported that there had been a positive effect of economical liberalisation in the stabilising of the macro-economic environment at the end of 1997, and in the stemming of budgets. Economical liberalisation has, however, had overwhelmingly negative effects on the people and the environment. Poverty, malnutrition and mortality rates in the country have increased since 1991.

Mortality rates are closely linked to nutritional status. According to the (British) Department for International Development (DFID) *Strategy Paper for Zambia* (1999), the nutritional status of the children in the country has deteriorated since 1991. Forty-two per cent of children under five are stunted (chronically malnourished), 18 per cent are severely stunted and up to 50 per cent suffer from micro-nutrient deficiencies. This has been exacerbated by poor access to safe water and sanitation. While the urban and peri-urban have piped water, water supply to many rural communities is usually from wells and boreholes or springs. In many cases, it is from (untreated) streams or through riverbank infiltration (ECZ, 2001). Poverty in Zambia is acute. The DFID Strategy Paper reports:

> Eighty-five per cent of the population lives on less than US$1 a day. Inequality is high: in 1998 per capita income in Zambia was about US$380, having fallen by some 60 per cent between 1975 and 1995. Zambia is highly urbanised by regional standards, with just under half the population living in urban areas. Declining incomes led to worsening poverty, but since 1995 there has been a very slight reduction in both the incidence and depth of national and rural (though not urban) poverty. Rural poverty remains more widespread and more severe than urban poverty (DFID, 1999).

Although the cause-effect relationship between poverty and environmental degradation has been widely acknowledged since the publication of the report of the World Commission on Sustainable Development in 1987, the concept of poverty is complex and so is its link to environmental degradation. The link between poverty and environment is characterised as a "vicious circle" or a "downward spiral" (ECZ, 2001:12). The *Zambia Human Development Report* of December 1998 (UNDP, 1998) indicates that there is a strong correlation between poverty and environmental degradation due to people's dependency on the exploitation of natural resources for their survival. The people of Chief Chiundaponde in Mpika district, for example, live in abject poverty. This has led to the plundering of the animal resources in the area. The nearby Lavushi Manda National Park has been depleted of animals through poaching. Efforts by the National Parks and Wildlife Services (now called Zambia Wildlife Authority) to curb rampant poaching resulted in resentment and conflicts with the local people. Hostility between them and wildlife police officers was rife. The law enforcers, tour operators and safari hunters who visited the area were perceived by the local people to be outsiders who came to plunder the wildlife. Since WWF Zambia Education Project entered the area with the concept of environmental education four years ago, the hostility is now in the past. We have used education as a social democratisation strategy that has allowed more people to take part in debates affecting their livelihoods. Through environmental education processes we have helped to resolve the conflicts, to build consensus and harmony between the community members and other stakeholders in the area. A sense of goodwill towards WWF and other people working in the area has been established.

In order to address issues associated with environmental degradation, the Zambian government has embarked on a number of sector reform programmes such as the Zambia Social Investment Fund (ZAMSIF), Environmental Support Programme (ESP), Agriculture Sector Investment Programme (ASIP) and many others falling under various ministries such as Tourism, Environment and Natural Resources, Agriculture, Food and Fisheries. This effort is further complemented by various laws and regulations that have been formulated to address environmental matters.

The shaping of environmental policies and institutions in Zambia

Following governance models introduced through colonialism, Zambia has formulated many laws and regulations addressing environmental matters since the colonial days. These include one environmental law, eight resource conservation laws and at least 24 laws of an administrative nature. The country has also enacted several statutory instruments that amplify the provisions of various acts. Although some laws have been derived from sectoral policies (Forestry, Wildlife and Water), the country has no national policy on the environment or environmental education (ECZ, 2001:141). Laws to stop or reduce environmental degradation have been inadequate as these are administered at sectoral level with different government departments working in isolation. They do not provide for an embracing natural resource conservation strategy, which would encourage voluntary compliance and enhance sustainable development. Many of these policies and strategies have been influenced by international policy developments guiding environmental change, and almost all of them have influenced the work that we do as WWF Zambia.

In 1985, five years after the publication of the *World Conservation Strategy* (Carwardine, 1987), the Zambian government adopted the *National Conservation Strategy* (NCS) (Government of the Republic of Zambia, 1985) as the principal policy that would guide sustainable use of the country's natural resources. The focus of the strategy was on the central role natural resources play in enhancing development. This saw the creation of the Ministry of Environment and Natural Resources in 1991. Following the Earth Summit in Rio de Janeiro in June 1992, Zambia endorsed the Agenda 21 Plan of Action and other Rio agreements. The Zambian government, with assistance from the UNDP, the World Bank and the Norwegian Agency for Development Co-operation (NORAD), developed the National Environment Action Plan (NEAP) that was adopted in 1994. The NEAP identifies five main areas of environmental concern in Zambia as:

- wildlife depletion;
- air pollution;
- land degradation;
- deforestation;
- water pollution and inadequate sanitation.

The NEAP's implementation involved the development of the Environmental Support Programme (ESP). This investment plan addressed issues concerning natural resource management activities and conservation, sustainable economic planning, public awareness, environmental information systems, research and policy development. I have been involved in the ESP programmes, on behalf of WWF ZEP, in facilitating some of the learning processes with communities as well as with teachers. As a result of ESP the project I work for has been able to reach many people in the rural areas through workshops organised by the programme. In fact the programme has similar activities to WWF ZEP in as far as environmental education is concerned, and this has helped our project to concentrate on small geographical areas since other far-flung areas are covered by ESP projects.

The Environmental Protection and Pollution Control Act (EPPCA) was enacted in 1990 as a further development of the NCS. This paved the way for the establishment of the Environmental Council of Zambia (ECZ) in 1992. WWF Zambia has been complementing the efforts of ECZ. It sits on the ECZ steering committee on wetlands policy formulation and, besides running its own wetland projects, it offers financial and technical support to ECZ wetland programmes in the Lukanga Swamps. Being the focal institution of the SADC[4] Regional Environmental Education Programme, ECZ has helped ZEP access part funding for running training workshops in environmental education. The council helps ZEP refocus its activities in line with government policies.

Other laws include the Zambia Wildlife Authority Act of 1998 and the Forestry Act of 1999. These and many other laws respond to the global development aid demands for community involvement in the management and utilisation of natural resources. As noted above, global developments have substantially influenced policy-making, and the socio-economic profile of Zambian society. Environmental policy-making, in particular, has been widely influenced by international environmental policies, treaties and global developments. From the above description it is clear, too, that all of these developments have an influence on practitioners like myself, who constantly need to keep up with new policies and changes in structures. Many of these developments have had positive outcomes (e.g. greater attention is being paid to the issues associated with deforestation) and many structures such as the Environmental Council of Zambia and Environmental Support Programme have been put in place to respond to environmental issues and risks. It is, however, not only at the level of policy, economics and structure that global influences have shaped practice in a local context.

Shaping environmental education in Zambia

In the early 1990s, a new radical view of environmental education, with a focus on social change and critique, emerged in southern Africa. This perspective was influenced by the socially critical orientation to environmental education, which gained currency in Australia through the works of Fien (1993), Robottom (1987), Greenall Gough & Robottom (1993) and many others. These environmental educators were questioning the institutional context and scientific roots of environmental education, which had dominated the understanding of the concept (and associated practice) prior to 1990.

Educators in southern Africa (particularly South Africans) were influenced by the work of these and other socially critical theorists, particularly since these ideas resonated with the social and political transformations that swept across the region in the early 1990s. Critical theorists have influenced my own practice as an environmental educator. In my earlier practice, I worked primarily with a scientific bias to environmental education. In recent years, I have worked more closely within a social focus, encouraging communities to participate in activities. I also draw on their prior knowledge and experience in my educational programmes, and I pay attention to the way in which others ascribe meaning to situations and contexts.

Although NGOs such as WWF ZEP have made headway in conceptualising environmental education, drawing on these regional and global trends by running comprehensive courses in environmental education, the Zambian government is still grappling with the concept. Its major focus is on the geographical or scientific approach. There is still a lot of uncertainty as to the distinction between environmental education and public awareness (ECZ, 2001)

Environmental educators (Fien, 1993; Huckle 1991; Lotz 1996) have argued that environmental education is more than knowledge provision and awareness-raising. Lotz (1999:51) describe environmental education as "a range of educational processes through which we respond to environmental issues in order to foster change in the direction of sustainable community life in a healthy environment". This focus on fostering social change has become an important dimension of our work at WWF ZEP over the past few years.

The shaping of indigenous environmental education practices

Changes taking place in the global arena have shifted the frameworks and processes of decision-making with regards to conservation and management of natural resources. Before colonialism, decisions associated with the conservation of natural resources were made through traditional social structures. Decision-making is now shaped and influenced by global trends and developments.

For the vast majority of African societies, education was an integral part of everyday life, long before the colonialists and western educationists came to Africa (Mwanakatwe, 1968; WWF ZEP, 1999). Traditional education involved a detailed understanding of the local biological resources and helped to develop knowledge and skills that enabled people to adapt to, and manipulate, their land, flora and fauna. Mwanakatwe further observes that by the time a child becomes a teenager, he/she has been exposed to an educational process enabling the acquisition of knowledge and skills for survival and adulthood. In traditional education, conservation was often realised in a pattern of shared beliefs, cultural taboos, folklore and myths. These embodied a common interest among communities to conserve natural resources.

Before the 1930 Native Authority Act in Zambia, natural resources were under the custodianship of traditional leaders. Chiefs controlled and imposed utilisation restrictions for the common good of every community member. The Unga chiefs of the Lake Bangweulu fishing communities in Luapula Province, for example, imposed a fishing ban (Banda, 1998) at certain times of the year (which coincides with a modern fishing ban under the Fisheries Act). During the ban, no one was allowed to fish (WWF ZEP, 1999). The Ila people of Mumbwa district shift their cattle from the villages to the Kafue flats during the dry season, in search of abundant greener pastures. The seasonal transfer of animals reduces pressure on pastures near settlements and improves the nutritional status of the animals. In the Western Province, Paramount Chief Litunga of the Lozi-speaking people was the sole custodian of rules and regulations governing hunting and exploitation of resources. He authorised when to hunt and gave hunting quotas (similar to those issued under the present Zambia Wildlife Authority Act of 1998). The Indunas (the chief's assistants) ensured that every member of the Lozi Kingdom adhered to the conservation rules and regulations. In Chieftainess Chiawa of the Goba-speaking people's area, big game such as hippos and elephants were only killed under special permission from the chief. The elderly people of Chiawa have told me that when a big animal was killed, the meat was shared among all the villages and was never sold for cash.

The WWF environmental education manual for teacher-educators (WWF ZEP, 1999:53–55) outlines some of these traditional conservation practices in terms of food security, natural resource management techniques, cultivation systems, health provision and domestication of animals. Traditional education and conservation systems were dismantled by colonisation and the emergence of scientific and conservation organisations such as the International Union for the Conservation of Nature and Natural Resources (IUCN) and WWF whose sole purpose was conservation of wildlife resources. This may have been justified by the fact that there was rampant depletion of animal resources at the time. Nevertheless, the initial conservation ideas brought in by conservation organisations tended to ignore the social impacts of the creation of national parks and game reserves, as the importance of nature conservation was emphasised through new structures and approaches to managing wildlife and natural resources in Africa. With these developments, "environment" was mostly seen in terms of nature (Thompson, 1998).

At the dawn of the 1970s, there were many international conferences, commissions and publications debating the issue of conservation. These debates have come to inform the way we view the environment. The *World Conservation Strategy* (IUCN *et al.*, 1980), "based on the views of more than 700 scientists and 450 government agencies from 100 countries" (Carwardine, 1987:67), stresses the importance of conservation and sustainable utilisation of life support systems, biodiversity and renewable resources. It identifies the priorities of conservation and presents an agreed-upon approach and broad plan for achieving them. Every country is mandated to come up with its own strategy, which looks at priority conservation areas, stimulates the appropriate action, raises public awareness and works to overcome any obstacles to taking action. As already mentioned, Zambia became the first African country to launch the *National Conservation Strategy* in 1985. The long-term goal of the NCS for Zambia is to satisfy the basic needs of all its nationals through the conservation and wise management of natural resources. "The strategy works on the principle that conservation and development are two sides of the same coin. Conservation can aid development by managing the natural resources on which it depends. Development can aid conservation by ensuring that people are not obliged to over-exploit natural resources in order to survive" (Carwardine, 1987:67).

More international debate saw the importance of broadening the view of environmental problems from the scientists' perspectives to include those of non-scientists across the globe. The World Commission on Environment and Development (WCED), under the aegis of the United Nations, published the Brundtland Report, *Our Common Future* (WCED, 1987). This publication opened up debate on the concept of sustainable development. As a follow-up to the *World Conservation Strategy*, *Caring for the Earth* (IUCN *et al.*, 1991) broadened the solution to environmental crisis by introducing the term "sustainable living". This brought another chance of re-examining conservation and development as it was done at the NGO Forum during the Earth Summit in 1992. The above organisations and commissions have all recommended a number of responses to environmental crises. One of the common features of these recommendations has been the need for education and training:

> Education, including formal education, public awareness and training, should be recognised as a process by which human beings and societies can reach their fullest potential. Education is critical for promoting sustainable development and improving the capacity of people to address environment and development issues (UNCED, 1992).

Zambian policy-makers have been considering the influence of these developments on environmental education practice in Zambia. The *State of the Environment Report* (ECZ, 2001:157), for example, notes that there has been a lot of progress made by different institutions to raise awareness, in light of the global ideas, at the level of environmental issues faced by the country. In June 2001, the Environmental Education and Public Awareness (EEPA) agency worked in collaboration with WWF ZEP and trained a number of community-based and non-governmental organisations in environmental education. This included sessions on community programme design, implementation, resource materials production and dissemination (Nalwamba & Lupele, 2001). In addition, EEPA and WWF ZEP have been planning a process of formulating a National Environmental Education Strategy. Thus, from this description, one can see the shift in focus from earlier traditional forms of environmental education, which were part of everyday life, to stronger institutional forms of environmental education, guided by international developments, and mediated by national and international organisations (see also O'Donoghue and Neluvhalani, this volume). Ironically, it is these global and regional perspectives on environmental education that are now being used to enable communities to make more sustainable choices in the context of their everyday lives, as I outlined in the opening description of WWF ZEP's current focus.

The shaping of Community-based Natural Resources management

When the law to create national parks and game reserves was enacted in the 1950s, people were relocated to make way for the creation of these protected areas and were thus alienated from the management and utilisation of natural resources (see also Moore and Masuku van Damme, this volume). In response, there has been a trend towards the Community-based Natural Resource Management (CBNRM) in many parts of Africa (see Booth *et al.*, 1997). The development of CBNRM has been strongly influenced by development aid programmes, as donors and governments sought ways of encouraging people to live in ways that promote sustainable development, especially where people are heavily dependent on the natural resource base for their livelihoods. In some cases this has led to real benefits for local communities. For example, the people in Lupande Game Management Area have built a school and a clinic from the proceeds of the South Luangwa Area Management Unit, a CBNRM project near the South Luangwa National Park (Chundama, Hachileka & Mupimpila, 1999). Some Game Management Areas have improved their economic status: "Mumbwa Game Management Area (GMA) receives significant income from the safari industry. It represents the fifth-highest revenue-earner of all GMAs and in Chibuluma village in particular, residents have a good understanding of the Administrative Management Design (ADMADE) programme and place a high value on wildlife" (Tilley, 1995:5).

However, as can be seen from several examples above, most of the global influences on the local are ambivalent. They not only have positive outcomes, but also are often difficult to implement, and more complex than initially assumed. In Zambia a CBNRM programme, ADMADE, was created by the government through the National Parks and Wildlife Services to help resolve some of the problems associated with conservation work as a result of the creation of game reserves and national parks. In this CBNRM initiative, Game Management Areas were established, with a view to enabling communities living near the reserves to share in the benefits of wildlife, and thus to encourage them to help manage wildlife as a valuable natural resource. ADMADE raised people's expectations. Local people were promised that 35 per cent of revenue from hunting concessions would be given to the communities in the GMA where the income was realised. In many cases the money raised was not enough for any meaningful community projects. In other cases the money was never given to the communities. After almost ten years of implementing the ADMADE programme, most communities in the GMAs where the programme was implemented still feel cheated (Simasiku, 2002). Chundama *et al.* (1998) observe that the incentives from wildlife in protected area management rarely trickle down to the community members. This makes understanding of the need for conservation difficult on the part of the people who are promised incentives when they participate in the conservation programmes. In addition, there are problems associated with the way in which the traditional leadership is seen to benefit more from wildlife resources in most CBNRM projects than are their subjects (Shackleton & Campbell, 2001; Simasiku, 2002). This has brought about some misunderstandings and mistrust among community members.

The voice of civil society

An area of importance to environmental education practitioners working in environments which are influenced by the ambivalence of global and local environmental ideas, policies and interventions is an understanding of how people are coming to understand and react to these influences. According to Zerner (2000) one of the offshoots of the political changes of the early 1990s in the southern African region has been the unmuzzling and amplification of historically silenced voices in the struggle for customary rights of access. The transformation from one-party segregative politics to more liberal multi-party politics has reconfigured political landscapes in Zambia, Tanzania, South Africa and Malawi. More people have been empowered through the work of civil society and NGOs to voice their concerns on issues affecting their livelihood (Zerner, 2000).

For example, eight traditional leaders in Zambia recently placed an advertisement in a local newspaper, explaining how they were ignored and neglected in the management of wildlife. They complained about poverty levels among the people in their chiefdoms despite the abundant wildlife resources in these areas. The advert read in part:

> We are now aware of the damage that has been done and how much has been amassed to the outsiders at the expense of the rural masses. We are aware that the poverty in which we live is not of our doing but has been done to us by a succession of uncaring and deliberately misinforming governments (*The Post*, 26 February 2002).

This awakening of community groups has not spared NGOs that are working in rural communities. The people of Chief Mwanachingwala in Mazubaka district obtained a court order to stop a WWF Partners for Wetlands Project from establishing a conservancy. They claimed that they were not consulted and they felt cheated by WWF and its business partners. The people were worried that their grazing land would eventually be grabbed if they allowed 47 000 hectares of their land to be fenced off:

> The Lusaka High Court has granted an interim injunction in favour of 700 villages in chief Mwanachingwala's area against WWF over proposals to fence off 47 000 hectares of pastoral land for a proposed conservation zone. According to the affidavit signed by one Pat Mwambwa on behalf of 700 others the action comes as a result of a petition against the proposed project being jointly handled by WWF and Mwanachingwala Conservation Limited. Mwambwa contends that being from an area and a subsistence farming village heavily reliant on pastoral activities, he was worried that there was going to be an increased level of skirmishes within the village over the grazing of animals (*The Post*, 3 April 2002, p.4).

There is increased response from civil society to the ambivalent influence of global ideas, policies and interventions even in contexts where poverty levels are high, and communities have few options in terms of making decisions about their livelihoods (see also Babikwa, 2002, this volume). I have found that community protest and situations of conflict are increasingly characterising the nature of our work. For example, our work in Chieftainess Chiawa's area was slowed down because of the conflicts and mistrusts between the community and Zambia Wildlife Authority (ZAWA). The latter is accused of pocketing money from tour operators' levies meant for community development activities. Since some of the ZAWA vehicles were bought by WWF funds and carry the organisation's logo, community members have difficulty in differentiating the two and sometimes mistake us for ZAWA officers.

Funding patterns and WWF ZEP initiatives

WWF International established WWF ZEP in 1989. Along the way, the project changed hands, for funding purposes, from WWF International to WWF UK under a joint funding scheme with the Department for International Development (DFID) of the British government. The project focus shifted with the donor demands. While it was more concerned with the provision of education materials for the lower basic primary school level in the first phase, the shift in the second phase was towards poverty alleviation in selected communities, in line with DFID interests (WWF ZEP, 1995). Community environmental education was further necessitated by the need to develop an integrated strategic plan for WWF in Zambia and the southern African region as a whole. One of the key strategic priorities identified was the designing of Community-based Natural Resource Management Systems. These developments have influenced WWF ZEP's current goals:

> To increase capacities in rural communities in Zambia to address their environment and development concerns and to understand, initiate and participate in the sustainable utilisation and management of their natural resources for short- and long-term benefits aimed at alleviating poverty (WWF ZEP, 2000).

WWF ZEP therefore established community projects in Mumbwa and Mpika in the Central and Northern Provinces respectively. Bee keeping and fish farming were just two of the community projects decided upon after involving the community members in environmental education processes such as Participatory Rural Appraisal. Being aware of the need for enabling processes that allow community voices and community perspectives, we were able to influence the project design, and to negotiate the boundaries of the project within the broader DFID mandate, available funding and the logical framework of planning restrictions. For example, during the funding proposal writing for the third phase, we made it clear that the local people would develop resource materials locally, in a participatory manner. We further indicated that the materials would be issue- and context-specific. Although we found it convincing, for funding purposes, to talk of participatory materials development and the use of action research, we had very scanty ideas of what exactly was involved. However, planning by the logframe global planning tool helped us to disguise our lack of competence in the matter. We were able to come up with indicators, objectives and assumptions, even when these were not shared with the people we were suppose to work with (Feuerstein, 1996). Logframes do not engage with or reflect the changing socio-historical issues and contextual changes in the project milieu. This once again illustrates the ambivalence of global influences on the local – the project direction was "constrained" by the project design process, and by donor priorities, yet we were able to establish an important process focus within this. While being able to do this, we nevertheless drew on the "security" of the project design product (the logical framework), and in doing so, we were not as "participatory" as we had intended to be. What was important in coming to terms with this ambivalence, however, was our developing experience of working in community-based contexts, around the priority focus areas we were trying to establish.

Participatory materials development work in WWF ZEP

In 1998, WWF ZEP was contracted by the Lake Tanganyika Biodiversity Project (LTBP) to develop posters for schools in the Lake Tanganyika catchment area in Mpulungu. This project was aimed at responding to issues of environmental degradation in line with Zambian environmental policy intentions. Influenced by international and regional ideas in environmental education, particularly the ideas associated with socially critical environmental education processes, WWF ZEP grappled with some of the issues associated with participatory materials development. We commissioned two artists, an editor, a curriculum specialist and 12 teachers from the 13 schools in the catchment to participate in a five-day poster-production workshop. They participated throughout the designing, information gathering, and final development of the scope and sequence chart stages. The posters produced were all problem-solving in orientation, highlighting issues that affected the catchment area. Captions were provided to consolidate the effectiveness of some of the posters. This was the first comprehensive attempt at participatory materials development processes in the WWF ZEP (Lupele, 2000). Besides drawing on the influence of international educational ideas, this process was also developed in response to assumptions that influence development work in a global context which indicates that one of the ways of ensuring sustainability of a project is to allow the participation of the beneficiaries (see Mikkelsen, 1995).

While the materials development process was successful, and high-quality materials were developed with the participation of communities, these materials were never produced further than camera-ready state. This was because the materials were produced towards the end of the LTBP life, with the hope of its renewal, illuminating a further dimension of global-local ambivalence. Projects often end, unfinished, after raising a lot of expectations among the local people.

This experience has, however, led to further participatory approaches to materials development in the WWF ZEP. We have established two participatory materials development projects, one in the Chiawa community and another in the Nalusanga community. To better understand the contextual complexities, the people we are working with, their history, the history of the area, and the current factors influencing their livelihood strategies, I have developed two contextual profiles, to inform us as we develop the materials in these contexts. We have found this to be a meaningful way of considering the local context, as well as the global influences on our work (see Lupele, 2002). The materials have been produced, and they are currently being used to enable communities to consider different options for their own development. In using the materials, people in these two communities are considering different options, and considering issues associated with longer-term impacts of unsustainable natural resource use.

Conclusion

Global factors play a very significant role in shaping what happens in the southern African context. Environmental education practices are influenced through policy changes, educational ideas and concepts that influence the way we think about our practice, funding patterns, models and approaches to development. These influences are of an ambivalent nature. They are both positive and negative, and often the same global influence has both positive and negative impacts on the local scene. However, local voices and practices are able to influence and shape these global forces in contextually significant ways. Thus, as deliberations in the aftermath of the United Nations World Summit on Sustainable Development begin to shape our local practices, it would be useful to bear in mind the complexity of local contexts and the unpredictable ways in which they are influenced.

One small strategy of trying to better understand the dynamic nature of the local context is to develop a contextual profile of the area one is working in (Lotz-Sisitka & Janse van Rensburg, 2000). Contextual profiles provide insights into the ongoing changes in the project milieu, for example the changing trends in resource management and utilisation. The local people's view of their practice and what should be done often provides a good starting point. This approach has worked well for me as I work to produce environmental education materials with community members. A contextual profile can supplement the logical framework, enabling us to consider not only the local influences on the project, but also the global influences, and how these shape a project and its direction.

Endnotes

1 A sub-species of antelope.
2 Richard Jeffery, Former Country Representative, WWF Zambia Country Office. Personal communication, 14 December 2001.
3 A name for the millipede in most Zambian vernaculars.
4 Southern African Development Community.

Achterberg, W. 1994a. *Samenleving, Natuur en Duurzaamheid. Een Inleiding in de Milieufilosofie.* Assen: Van Gorcum.

Achterberg, W. 1994b. Can liberal democracy survive the environmental crisis? Sustainability, liberal neutrality and overlapping consensus. In: Zweers, W. & Boersema, J.J. (eds). *Ecology, Technology and Culture. Essays in Environmental Philosophy.* Cambridge: The White Horse Press.

Achterhuis, H. 1994. The lie of sustainability. In: Zweers, W. & Boersema, J.J. (eds). *Ecology, Technology and Culture. Essays in Environmental Philosophy.* Cambridge: The White Horse Press.

African National Congress (ANC). 1994. *A Policy Framework for Education and Training: Discussion Document.* Centre for Education Policy Development/SACHED, Braamfontein.

Anderson, D. & Grove, R. 1993. *Conservation in Africa. People, Policies and Practice.* Cambridge: Cambridge University Press.

Anderson, V. & Johnson, L. 1997. *Systems Thinking Basics: From Concepts to Causal Loops.* Waltham: Pegasus Communications.

Asmal, K. 1999. *Call to Action: Mobilizing Citizens to Build a South African Education System for the 21st Century.* Pretoria: Department of Education.

Baden-Powell, R.S.S. 1897. *The Matabele Campaign.* London: Methuen.

Baden-Powell, R.S.S. 1900. *Sport in War.* London: Heinemann.

Banda, M. 1998. Collaborative Fish Management. Paper presented at the Trialing Workshop for the Fish Conservation Book (Second Edition) held by WWF Zambia Education Project in Samfya, Zambia.

Bauman, Z. 1993. *Postmodern Ethics.* Oxford: Blackwell.

Bauman, Z. 1995. *Life in Fragments: Essays in Postmodern Morality.* Oxford: Blackwell.

Bauman, Z. 2000. *Liquid Modernity.* Cambridge: Polity Press.

Bauman, Z. 2001. *The Individualized Society.* Cambridge: Polity Press.

Bauman, Z. 1987. *Legislators and Interpreters – On Modernity, Post-modernity and Intellectuals.* Cambridge: Polity Press.

Beck, U. 1992. *Risk Society: Towards a New Modernity.* London: Sage.

Beck, U. 1999. *World Risk Society.* London: Blackwell.

Beilharz, P. 2001. *The Bauman Reader.* Oxford: Blackwell.

Berger, G. 1996. The Internet. Goldmine for Journalists. *Rhodes Journalism Review*, Number 12. Department of Journalism and Media Studies, Rhodes University, Grahamstown.

Beyene, A. 2000 Decentralization as a tool for resolving the nationality problem: The Ethiopian experience. *Regional Development Dialogue*, 21(1):122–147.

Bhaskar, R. 1989. *Reclaiming Reality: A Critical Introduction to Contemporary Philosophy.* London: Verso.

Binns, T. (ed.) 1995. *People and Environment in Africa.* Chichester, England: John Wiley & Sons.

Bjerg-Hansen, K. & Sguazzin, T. 2000. *Sustainable Stories.* Johannesburg: Learning for Sustainability Project.

Blowers, A. 1997. Environmental policy: Ecological modernisation or the risk society. *Urban Studies*, 43(5–6):845–871.

Booth, A., McCullum, J., Mpinga, J. & Mukute, M. 1997. *State of the Environment in Southern Africa.* Harare, Zimbabwe: IUCN.

Bossuyt, J. & Gould, J. 2000. Decentralisation and poverty reduction: Elaborating the linkages. *Policy Management Brief*, 12. European Centre for Development Policy Management, Maastricht.

Bourdieu, P. & Passeron, J. 1977. *Reproduction in Education, Society and Culture.* London: Sage.

Bourdieu, P. 1998. *Practical Reason: On the Theory of Action.* Cambridge: Polity Press.

Brazier, C. 1999. Making it Happen. The Great Education Scandal. *New Internationalist*, No. 315.

Bromley, D.M. & Cernea, M.M. 1989. The management of common property natural resources: Some conceptual and operational fallacies. *World Bank Discussion Papers*, 57. World Bank, Washington, DC.

Burbules, N. & Torres, C. 2000. *Globalisation and Education: Critical Perspectives.* New York: Routledge.

Carr, W. & Kemmis, S. 1986. *Becoming Critical: Education Knowledge and Action Research.* London: Falmer Press, and Geelong, Australia: Deakin University Press.

Carruthers, J. 1995. *The Kruger National Park: A Social and Political History.* Pietermaritzburg , University of Natal Press.

Carwardine, M. 1987. *The Nature of Zambia: A Guide to Conservation and Development Issues No. 2.* IUCN, Gland, Switzerland.

Casalis, E. 1861. *The Basutos.* Morija: Morija Printing Works.

Castoriadis, C. 1981. From ecology to autonomy. *Thesis Eleven*, 3:7–22.

Cheema, G.S. & Rondinelli, D.A. (eds). 1983. *Decentralization and Development: Policy Implementation in Developing Countries.* Beverly Hills: Sage.

Chinweizu. 1987. *Decolonising the African Mind.* Nigeria: Pero Press.

Chundama, M., Hachileka, E. & Mupimpila, C. 1999. *The Effectiveness of Benefit Sharing Schemes in Community-Based Wildlife Resource Management Programmes in Zambia.* Unpublished Special Study, Ministry of Finance and Economic Development. Lusaka, Zambia.

Clacherty, A. 1993a. The Environmental Education Policy Initiative: Reflections of the process. *Southern African Journal of Environmental Education*, 13: 3–6.

Clacherty, A. 1993b. *The Incorporation of Environmental Education into Formal Education.* Final report presented to the Department of Environment Affairs, 30 November. Johannesburg: EEPI.

Clacherty, A. 1995. *Environmental Education Policy Options for Formal Education in South Africa: A Source Document for Curriculum Development in Environmental Education.* Johannesburg: EEPI.

Cock, J. & Fig, D. 2002. From Colonial to Community-based Conservation. In: McDonald, D.A. (ed.). *Environmental Justice in South Africa.* Cape Town: UCT Press.

Committee for the Review of Curriculum 2005. 2000. *A South African Curriculum for the Twenty-First Century.* Report Presented to the Minister of Education. Pretoria: Department of Education.

Converey, F.J. 1992. *The Economic Aspects of the National Environmental Action Plan Process in Uganda.* Dublin: University College of Dublin Press.

Cornbleth, C. 1991. *Curriculum in Context.* London: Falmer Press.

Crook, R.C. & Manor, J. 1998. *Democracy and Decentralization in South-East Asia and West Africa: Participation, Accountability and Performance.* Cambridge: Cambridge University Press.

Cross, M. 1999. *Imagery and Identity in South African Education, 1880–1990.* North Carolina: Carolina Academic Press.

Crossman, P. 2002. (ed.). *Teaching Endogenous Knowledge in South Africa: Issues, Approaches and Aids.* Pretoria: University of Pretoria.

Czerniewicz, L., Murray, S. & Probyn, M. 2000. *The Role of Learning Support Materials in C2005.* Pretoria: National Centre for Curriculum Research and Development.

Daly, H.E. (ed.). 1973. *Toward a Steady-State Economy.* San Francisco: Freeman.

Day, B. & Monroe, M. (eds). 2000. *Environmental Education & Communication for a Sustainable World – Handbook for International Practitioners.* Washington, DC: Academy for Educational Development.

Department for International Development, Great Britain (DFID). 1999. *Strategy Paper for Zambia.* London: DFID.

Department of Constitutional Development (DCD). 1998. *White Paper for Local Government.* Government Notice No 18739 of 1998. Pretoria: Government Printers.

Department of Education (DoE). 1995. *A Curriculum Framework for General and Further Education and Training.* Discussion document developed by the Consultative Forum on Curriculum. December. Pretoria: DoE.

Department of Education (DoE). 1996. *Draft Curriculum Framework Document.* Pretoria: DoE.

Department of Education (DoE). 1997. Curriculum Policy Documents. Pretoria: DoE.

Department of Education (DoE). 1998. *Guidelines for the Development of Learning Support Materials.* Pretoria: DoE.

Department of Education (DoE). 2000. *An Introduction to the Theory and Practice of Outcomes-Based Education in South Africa.* Training material for Grade 4 and Grade 8 National Pilot. Warmbaths, 13–24 March. Pretoria: DoE.

Department of Education (DoE). 2002. *Revised National Curriculum Statement. Grades R–9 (Schools).* Pretoria: DoE.

Department of Education (DoE) & Department of Labour (DoL). 1995. *White Paper on Education and Training.* Government Gazette, Notice No. 196 of 1995. Cape Town: Government Printers.

Department of Education (DoE) & Department of Labour (DoL). 2002. *Report of the Study Team on the Implementation of the National Qualifications Framework.* Pretoria: DoE & DoL.

Department of Education (DoE) & EEASA. 2000. *EnviroDays 2000.* Howick, South Africa: Share-Net.

Department of Environmental Affairs. 1998. *State of Environment Report for Malawi.* Lilongwe: Department of Environmental Affairs.

Department of Environmental Affairs and Tourism (DEAT). 1999. *White Paper on Environmental Management Policy in South Africa.* Cape Town: DEAT.

Department of Research and Environmental Affairs. 1994. *The National Environmental Action Plan.* Lilongwe: Department of Environmental Affairs.

Dladla, Y. (ed.). 1998. *Voices, Values and Identities Symposium.* Pretoria: SANParks.

Doll, W. 1993. *A Post-Modern Perspective on Curriculum.* New York: Teachers College Press.

Dreyfus, A., Wals, A. & Van Weelie, D. 1999. Biodiversity as a postmodern theme for environmental education. *Canadian Journal of Environmental Education,* 4: 155–175.

Driver, T. 1998. *The Theory of Politics of Mountain Rangeland Conservation and Pastoral Development in Colonial Lesotho.* Unpublished doctoral thesis, University of London.

Du Toit, D. & Sguazzin, T. 1999. *Camera and Context: Using Camera to Explore Environment and Curriculum Development.* Johannesburg: Learning for Sustainability Project.

Durban City Council. 1998. *Durban's Local Agenda 21 Programme … the new way of doing things ….* Durban: Durban City Council.

Durban Solid Waste (DSW). 2000. *A Recycling Business Plan for the Durban Metropolitan Region for the Creation of Jobs.* Unpublished business plan. Durban: DSW.

Durban Solid Waste (DSW). 2002. Waste Minimisation Office. Introduction. Unpublished pamphlet. Durban: DSW.

Earth Policy Institute. 2001. Wind Power: The Missing Link in the Bush Energy Plan. *Earth Policy Alert,* 2001–3. http://www.earth-policy.org/Alerts/ Alert14.htm.

Eckersley, R. 1992 *Environmentalism and Political Theory. Towards an Ecocentric Approach.* London: UCL Press.

Edwards, D. 1998. Can we learn the truth about the environment from the media? *The Ecologist,* 28(1).

EEASA. 1992. EEASA, the workshop, some achievements and the challenges ahead. *Environmental Education and Development: Proceedings of the Tenth Anniversary Workshop, Howick, Natal: 16–18 July.* Howick: EEASA.

Eldredge, E. 1993. *A South African Kingdom.* Cambridge: Cambridge University Press.

Ellsworth, E. 1989. Why doesn't this feel empowering? Working through the repressive myths of critical pedagogy. *Harvard Educational Review,* 59(3):297–324.

Environmental Council of Zambia (ECZ). 1994. *State of the Environment in Zambia 1994.* Lusaka, Zambia: ECZ.

Environmental Council of Zambia (ECZ). 2001. *State of the Environment in Zambia 2000.* Lusaka, Zambia: ECZ.

ESKOM. 2000. *ESKOM Environmental Report, 2000.* http://www.eskom.co.za/enviroreport01/ index.htm.

ESKOM. 2001. *ESKOM Annual Report, 2001.* http://www.eskom.co.za/annreport02/index.htm.

Fairclough, N. 1992. *Discourse and Social Change.* Cambridge: Polity Press.

Fardon, R. & Furniss, G. 2000. *African Broadcast Cultures: Radio in Transition.* Cape Town: David Philip.

Feenberg, A. 1979. Beyond the politics of survival. *Theory and Society,* 7:319–361.

Ferguson, J. 1997. Development and Bureaucratic Power in Lesotho. In: Rahnema, M. & Bawtree, V. (eds). *The Post Development Reader.* London: University Press.

Feuerstein, M. 1986. *Partners in Evaluation: Evaluating Development and Community Programmes with Participants.* London: MacMillan Education.

Fien, J. 1993. *Education for the Environment: Critical Curriculum Theorising and Environmental Education.* Geelong, Australia: Deakin University Press.

Fien, J. 1996. Introduction to environmental education. *Teaching for a Sustainable World.* Geneva: UNESCO.

Financial Mail. 2000. SA National Parks: The Main Attraction. Supplement to *Financial Mail,* 3 November. Johannesburg.

Fishman, M. 1977. *Manufacturing the News.* California: University of Texas.

Foucault, M. 1975. *Discipline and Punish – The Birth of the Prison.* Harmondsworth: Penguin.

Fox, J. & Aranda, J. 1996. *Decentralization and Rural Development in Mexico: Community Participation in Oaxaca's Municipal Funds Program.* San Diego: Center for US-Mexican Studies, University of California.

Freeman, S. 2000. *The Role of Capacity Building in the Public Participation Process: The Case of Landfill Siting in the North of the Durban Metropolitan Area.* Unpublished Masters thesis, School of Life and Environmental Sciences, University of Natal-Durban, Durban.

Furtado, X. 2001. Decentralization and capacity development: Understanding the links and the implications for programming. *Capacity Development Occasional Paper Series CIDA Policy Branch,* No. 4. CIDA, Toronto.

Gare, A. 1995. *Postmodernism and the Environmental Crisis.* London: Routledge.

Gergen, K. 2001. *Social Construction in Context.* London: Sage.

Giddens, A. 1991. *Modernity and Self-Identity: Self and Society in the Late Modern Age.* Stanford, California: Stanford University Press.

Goldsmith, E., Allen, R., Allaby, M., Davoll, J. & Lawrence, S. 1972. *A Blueprint for Survival.* Harmondsworth: Penguin Books.

Goodson, I. 1994. *Studying Curriculum. Cases and Methods.* Buckingham: Open University Press.

Gough, A. 1997. *Education and the Environment: Policy, Trends and the Problems of Marginalization.* Melbourne: Australian Council for Educational Research.

Gough, N. 1998. Decolonising 'Sustainability': Environmental Education as a Postcolonial Textual Practice. In: *EEASA '98 Conference Proceedings, University of Botswana, 7–10 July.* Gaborone: EEASA.

Government of the Republic of Zambia, 1985. *The National Conservation Strategy for Zambia.* Gland, Switzerland: IUCN.

Greenall Gough, A. & Robottom, I. 1993. Towards a socially critical environmental education: Water quality studies in a coastal school. *Journal of Curriculum Studies*, 25 (4): 301–316.

Grove, R. 1993. Early Themes in African Conservation: The Cape in the Nineteenth Century. In: Anderson, D. & Grove, R. (eds). *Conservation in Africa. People, Policies and Practice.* Cambridge: Cambridge University Press.

Grundy, S. 1987. *Curriculum: Product or Praxis.* London: Falmer Press.

Habermas, J. 1970. Towards a theory of communicative competence. *Inquiry,* 13(4): 360–375.

Hachileka, E. 1997. *Natural Resources Economics.* The Human and Social Perspectives in Natural Resources Management Course. In-Service Training Trust, Lusaka, Zambia.

Hajer, M. 1997. *The Politics of Environmental Discourse – Ecological Modernization and the Policy Process.* New York: Oxford University Press.

Hannigan, J.A. 1995. *Environmental Sociology: A Social Constructionist Perspective.* London: Routledge.

Hannigan, J. 1999. *Environmental Sociology.* London: Routledge.

Haraldsson, H. 2002. *Course Notes. Introduction to Systems and Causal Loop Diagrams.* Environmental Science Masters Programme. Lund University, Sweden.

Hardin, G. 1968. The tragedy of the commons. *Science,* 162:1243–1248.

Hardin, G. 1992. The case against helping the poor. In: Bonevac, D. (ed.). *Today's Moral Issues. Classic and Contemporary Perspectives.* Mountain View, California: Mayfield.

Harley, K. & Parker, B. 1999. Integrating Differences: Implications of an Outcomes-based National Qualifications Framework for the Roles and Competencies of Teachers. In: Jansen, J. & Christie, P. (eds). *Changing Curriculum: Studies on Outcomes-Based Education in South Africa.* Kenwyn: Juta.

Hattingh, J.P. 2000. The history of sustainable development. *Report from the Unit for Environmental Ethics*, University of Stellenbosch. Special Issue, 1(1), Summer: 1–3.

Hattingh, J.P. 2002. Conceptualizing ecological sustainability and ecologically sustainable development in ethical terms: Issues and challenges. *Annale.* Stellenbosch: University of Stellenbosch.

Heilbroner, R.L. 1974. *An Inquiry into the Human Prospect.* New York: Norton.

Horrell, M. 1968. *Bantu Education to 1968.* Johannesburg: SA Institute of Race Relations.

Hugo, M.L. 1974. An investigation into the nature and magnitude of the demand for outdoor recreation among the urban black communities of South Africa. *The South African Journal of African Affairs*, 4(2).

International Council for Adult Education (ICAE). 1993. Treaty on environmental education for sustainable societies and global responsibility. *Adult Education and Development*, Vol. 40.

IUCN, UNEP, WWF. 1980. *World Conservation Strategy. Living Resource Conservation for Sustainable Development*. Gland, Switzerland: IUCN/UNEP/WWF.

IUCN, UNEP, WWF. 1991. *Caring for the Earth. A Strategy for Sustainable Living.* Gland, Switzerland: IUCN/UNEP/WWF.

IUCN. 1971. Educating and the environment. Papers of the Nevada Conference of 1970 and Zurich Conference of December 1971. *IUCN Publication New Series.* Morges: IUCN.

Jacobs, M. 1999. Sustainable development as a contested concept. In: Dobson, A. (ed.). *Fairness and Futurity. Essays on Environmental Sustainability and Social Justice.* Oxford: Oxford University Press.

Janks, H. 1997. Critical discourse analysis as a research tool. *Discourse Studies in the Cultural Politics of Education,* 18(3):329–342.

Janse van Rensburg, E. 1995. *Environmental Education and Research in Southern Africa: A Landscape of Shifting Priorities.* Ph.D thesis, Grahamstown, Rhodes University.

Janse van Rensburg, E. 2001. *Environmental Interpretation and Education for a Conservation Context.* Environmental Interpretation & Education Course, Strand 4. SANParks, Pretoria & Rhodes University, Grahamstown.

Janse van Rensburg, E. & Du Toit, D. 2000. 'Sustainability' from constructivist & socially critical angles: Ambiguous steering ideas. In: Janse van Rensburg, E. & Lotz-Sisitka, H. *Monograph: Learning for Sustainability.* Johannesburg: *Learning for Sustainability* Project.

Janse van Rensburg, E. & Lotz, H. 1998. *Enabling Environmental Education as a Cross-Curricular Concern in Outcomes-Based Learning Programmes.* Discussion Document, April. Howick, South Africa: Share-Net.

Janse van Rensburg, E. & Lotz-Sisitka, H. 2000. *Monograph: Learning for Sustainability.* Johannesburg: *Learning for Sustainability* Project.

Janse van Rensburg, E. & Mhoney, K. 2000. The spiral model of *Learning for Sustainability:* An exploration of process-oriented teacher development in environmental education. In: Janse van Rensburg, E. & Lotz-Sisitka, H. *Monograph: Learning for Sustainability.* Johannesburg: *Learning for Sustainability* Project.

Janse van Rensburg, E. & Van Harmelen, U. 1996. *Enabling Environmental Education in Outcomes-Based Education.* EECI Discussion Document. Rhodes University, Grahamstown.

Jansen, J. 1997. Why OBE will fail? Paper presented at the National Conference on Outcomes-Based Education, University of Durban-Westville, 20 March. Durban.

Jansen, J. 1999. Setting the scene: Historiographies of curriculum policy in South Africa. In: Jansen, J. & Christie, P. (eds). *Changing Curriculum: Studies on Outcomes-Based Education in South Africa.* Kenwyn: Juta.

Jansen, J. & Christie, P. (eds). 1999. *Changing Curriculum: Studies on Outcomes-Based Education in South Africa.* Kenwyn: Juta.

Jickling, B. 1999. Beyond sustainability: Should we expect more from education? *Southern African Journal of Environmental Education,* No. 19:60–67.

John, L. 2000. O poder da mídia na formação da agenda ambiental. Downloaded from http://www.jornalismoambiental.jor.br/artigos.htm on 23–28 June 2001.

Joubert, S.C.J. n.d. *Master Plan for the Management of the Kruger National Park.* Volumes 1 to 6. Mimeo. Kruger National Park: SANParks.

Katahoire A. & Ndidde, A. 1996. Participatory Evaluation in Action: Experiences of working with ADF funded projects in Uganda. Draft Manuscript. Kampala.

Kaunda, J. 2002. Villagers Stop WWF Deal. *The Post,* 3 April:4.

Keating, M. 1993. *Covering the Environment – A Handbook on Environmental Journalism*. Ontario: National Round Table on the Environment and the Economy.

Keegan, J. 1986. *Rural Transformation in Industrialised South Africa: The Southern Highveld to 1914*. Braamfontein, South Africa: Ravan Press.

Kemmis, S. & McTaggart, R. (eds). 1988. *The Action Research Planner*. Geelong, Australia: Deakin University Press.

Khan, F. 2002. The roots of environmental racism and the rise of environmental justice in the 1990s. In: McDonald, D.A. (ed.). *Environmental Justice in South Africa*. Cape Town: UCT Press.

Kincheloe, J. 1995. The struggle for a critical postmodern action research. In: McLaren, P. & Giarelli, J. (eds). *Critical Theory and Educational Research*. Albany: State University of New York Press.

Klein, C. 1997. *Participatory programme development at an environmental education center through action research involving secondary school teachers*. Unpublished Masters thesis, Grahamstown, Rhodes University.

Koch, E. 1998. Out There. *Ecofile*. March.

Kraak, A. 1999. Competing education and training discourses: A 'systematic' versus 'unit standards' framework. In: Jansen, J. & Christie, P. (eds). *Changing Curriculum: Studies on Outcomes-Based Education in South Africa*. Kenwyn: Juta.

Kress, G. & Hodge, R. 1988. *Social Semiotics*. New York: Cornell University Press.

Kyaddondo, D. & Kyomuhendo, S.S. 2000. *Food Security and Farm Enterprise Support Project: Baseline Survey Report*. Kampala: VEDCO.

Latour, B. 1993. An interview with Bruno Latour. Interviewed by T. Hugh Crawford. *Configurations*, 1(2):247–268. http://muse.jhu.edu/demo /configurations/ 1.2crawford.html.

Latour, B. 1999. *Pandora's Hope: Essays on the Reality of Science Studies*. Massachusetts: Harvard University Press.

Le Grange, L. & Reddy, C. 1997. Environmental Education and Outcomes-based Education in South Africa: A Marriage Made in Heaven? *Southern African Journal of Environmental Education*, 17:12–18.

Linney, B. 1995. *Pictures, People and Power*. London: MacMillan.

Local Government. 1998a. *District Development Planning Handbook*. Lilongwe: Local Government Decentralisation Secretariat.

Local Government. 1998b. *National Decentralisation Policy*. Lilongwe: Local Government Decentralisation Secretariat.

Local Government. 1998d. *Village Action Planning Manual*. Lilongwe: Local Government Decentralisation Secretariat.

Lotz, H. 1996. *The development of environmental education resource materials for Junior Primary Education through teacher participation: The case of the We Care Primary Project*. Unpublished D.Ed. thesis, University of Stellenbosch, Stellenbosch.

Lotz, H. (ed.). 1999. *Developing Curriculum Frameworks: A Source Book on Environmental Education Amongst Adult Learners*. Howick, South Africa: SADC Regional Environmental Education Centre.

Lotz, H. 1999a. *Learning for Sustainability Illustrated Report: School-based Learning Programme Unit Development with Teachers*. Grahamstown: Rhodes University Environmental Education Unit.

Lotz-Sisitka, H. & Ashwell, A. (n.y.p.). Policy and Praxis: Environmental Education Curriculum Development 1992–2002. Grahamstown: Rhodes University Environmental Education Unit.

Lotz-Sisitka, H. & Janse van Rensburg, E. 2000. *Learning for Sustainability Contextual Profile*. Johannesburg: Learning for Sustainability Project.

Lotz, H. & Olivier, C. 1998. Clarifying orientations to learning programme development within the Outcomes-based Education framework and the *Learning for Sustainability Project* in Gauteng and Mpumalanga. Paper presented at the Outcomes-Based International Symposium hosted by Vista University, 17–18 November. Pretoria.

Lotz-Sisitka, H. & Olivier, C. 2000. Teachers as curriculum developers: Insights from the *Learning for Sustainability Project*. In: Janse van Rensburg, E. & Lotz-Sisitka, H. *Monograph: Learning for Sustainability. Johannesburg: Learning for Sustainability* Project.

Lotz-Sisitka, H. & Raven, G. 2001. *Active Learning in OBE: Environmental Learning in South African Schools. Report of the National Environmental Education Programme – GET Pilot Research Project.* Pretoria: Department of Education.

Lotz, H. & Robottom, I. 1998. 'Environment as Text': Initial insights into Some Implications for Professional Development in Environmental Education. *Southern African Journal of Environmental Education*, 18:19–28.

Lotz, H., Tselane, T. & Wagiet, R. 1998. *Supporting Curriculum 2005. Developing Learning Programmes with Environment as Phase Organizer.* Pretoria: Department of Environmental Affairs and Tourism.

Lupele, J.K. 2000. Participatory materials development in Zambia: The workshop that worked. *EEASA Bulletin*, vol. 20.

Lupele, J.K. 2002. *Contextual Profile as a Monitoring Tool.* Unpublished Discussion Paper. WWF Zambia Education Project, Lusaka.

Machishi, P.C. & Musona, E.T. 1999. *The People of Zambia: A Short History of the Soli from 1500 to 1900.* Lusaka, Zambia: Multimedia Publications.

Mackenzie, J.M. 1993. Chivalry, social Darwinism and ritualised killing: The hunting ethos in Central Africa up to 1914. In: Anderson, D. & Grove, R. (eds). *Conservation in Africa. People, Policies and Practice.* Cambridge: Cambridge University Press.

Malawi Government. 1996. *Environmental Management Act.* Zomba: Government Printer.

Malawi Government. 1998. *Local Government Act.* Zomba: Government Printer.

Manda, P.H., Bwalya, T.M. & Simbeya, P.M. 1997. *Basic Education Project: Our World 6.* Lusaka: MacMillan Publishers.

Manqele, M. 1994. *Indigenous Knowledge Series.* Pilot Edition. Howick, South Africa: Share-Net.

Masuku, L. 1998. *Indigenous Knowledge Series.* Revised edition. Howick, South Africa: Share-Net.

Masuku, L. 1999. *The role of indigenous knowledge in/for environmental education: the case of an Nguni story in the schools water action project.* Unpublished Masters thesis, Rhodes University, Grahamstown.

McCarney, P.L. 1996. Reviving local government: the neglected tier in development. In: McCarney, P.L. (ed.). *The Changing Nature of Local Government in Developing Countries.* Toronto: Federation of Canadian Municipalities and the University of Toronto.

McKernan, J. 1991. *Curriculum Action Research: A Handbook of Methods and Resources for the Reflective Practitioner.* London: Kogan Page.

McKernan, J. 1993. Perspectives and imperatives: some limitations of outcomes-based education. *Journal of Curriculum and Supervision*, 8(4):343–353.

McTaggart, R. 1997. Guiding principles for participatory action research. In: McTaggart, R. (ed.). *Participatory Action Research. International Context and Consequences.* Albany: State University of New York Press.

Meadows, D.H., Meadows, D.L., Randers, J. & Behrens, W.W. (III). 1972. *The Limits to Growth. A Report for the Club of Rome's Project on the Predicament of Mankind.* New York: Universe Books.

Mies, M. & Shiva, V. 1993. *Ecofeminism.* London: Zed Books.

Mikkelsen, B. 1995. *Methods for Development Work and Research: A Guide for Practitioners*. New Delhi: Sage.

Mill, J.S. 1848. Principles of Political Economy. In: *Collected works of John Stuart Mill*, 1965, Vol. III. Toronto: University of Toronto Press.

Ministry of Agriculture, Animal Industry and Fisheries (MAAIF) & Ministry of Finance, Planning and Economic Development (MFPED). 2000. *Plan for Modernisation of Agriculture: Eradicating Poverty in Uganda*. Kampala: Government of Uganda.

Ministry of Environment and Natural Resources. 1994. *National Environmental Action Plan*. Lusaka: Ministry of Environment and Natural Resources.

Ministry of Natural Resources (MNR). 1995. *The National Environmental Action Plan*. Kampala: Government of Uganda.

Ministry of Research and Environmental Affairs (MOREA). 1995. *Environmental Support Programme*. Lilongwe: MOREA.

Ministry of Research and Environmental Affairs (MOREA). 1996. *National Environmental Policy*. Lilongwe: MOREA.

Mokuku, T. 1999. *Education for environmental literacy: Participatory action research in the science curriculum in Lesotho*. Unpublished Ph.D thesis, Grahamstown, Rhodes University.

Mokuku, T. 2001. Encounters with action research in the African context: A case study in the school curricula in Lesotho. *Educational Action Research: An International Journal*, 9(2):172–186.

Moll, I. 2001. Vygosky and Vygotsky-speak: Understanding the relationship between schooling and everyday knowledge. *Journal of Education*, 26:5-19.

Moll, I. 2002. Clarifying constructivism in a context of curriculum change. *Journal of Education*, 27:5–32.

Moore, K.T.B. 2002. *Social Ecology Survey of SANPARKS*. Pretoria: SANParks.

Morris, M. 1996. That hoary word ... *Rhodes Journalism Review*, Number 12. Department of Journalism and Media Studies, Rhodes University, Grahamstown.

Mshana, R. 1992. *Insisting Upon People's Knowledge to Resist Developmentalism: Peasant Communities as Producers of Knowledge for Social Transformation in Tanzania*. Verlag fur Interkulturelle Kommunikation, Frankfurt.

Mtshali, M. 1994. *An investigation of environmental knowledge among two rural black communities in Natal*. Unpublished Masters thesis, Rhodes University, Grahamstown.

Mulkay, M. 1991. *Sociology of Science*. Indiana University Press, Indianapolis.

Murombedzi, J. 1991. Decentralising common property resources management: A case study of the Nyaminyami District Council of Zimbabwe's Wildlife Management Programme. *International Institute for Environment and Development Paper 30*. London: IIED.

Murray, C. 1981. *Families Divided: The Impact of Migrant Labour in Lesotho*. Cambridge: Cambridge University Press.

Mwanakatwe, J.M. 1968. *The Growth of Education in Zambia Since Independence*. Lusaka, Zambia: Oxford University Press.

Nalwamba, C. & Lupele, J. 2001. *Training of Trainers Workshop for NGOs and CBOs in Community Environmental Management Programme (CEMP) and Community-Based Natural Resources Management (CBNRM) Districts*. Workshop Report, MENR-ESP, Kafue, Zambia.

National Education Policy Investigation (NEPI). 1993. *The National Education Policy Investigation. The Framework Report*. Cape Town: Oxford University Press.

National Environmental Education Project – General Education and Training (NEEP-GET). 2001a. *Project Document*. Pretoria: Danced/Department of Education.

National Environmental Education Project – General Education and Training. (NEEP-GET). 2001b. *Environment in the National Curriculum Statement. A Discussion Document for Learning Area Working Groups*. Pretoria: NEEP-GET.

Neluvhalani, E.F. 2002. Mobilising indigenous knowing in curriculum contexts of environmental learning: Case studies in the Limpopo province. Ph.D thesis proposal, Department of Education, Rhodes University, Grahamstown.

Ngugi wa Thiong'o. 1993. *Moving the Centre. The Struggle for Cultural Freedom*. Nairobi: EAEP.

Nogueira, A. 2001. Indymedia – wired dissent. *New Internationalist Magazine*, Number 333.

Norton, B. 1984. Environmental ethics and weak anthropocentrism. *Environmental Ethics*, 6 (Summer):131–148.

Norton, B. 1992. Sustainability, human welfare and ecosystem health. *Environmental Values*, 1(1):97–111.

Nthathakane, M. 2001. *Multicultural Science Education in Lesotho High School Biology Classrooms*. Ph.D thesis, University of Wisconsin, Madison.

O'Donoghue, R. 1993. *The Environment, Development and Environmental Education*. Howick: EEPI.

O'Donoghue, R. 1997. *Detached Harmonies: A Study in/on Developing Processes of Environmental Education in Eastern Southern Africa*. Unpublished PhD thesis, Rhodes University, Grahamstown.

O'Donoghue, R. 1999. Participation: An under theorised icon in research and curriculum development. *Southern African Journal of Environmental Education*, 19:14–27.

O'Donoghue, R. 2001a. *Environment and Active Learning in OBE. NEEP Guidelines for Facilitating and Assessing Active Learning in OBE*. Howick, South Africa: Share-Net.

O'Donoghue, R. 2001b. *The Environment and Active Learning Processes in Outcomes Based Education. National Environmental Education Programme (NEEP) Guidelines for Facilitating and Assessing Active Learning in OBE*. Howick, South Africa: Ministry of Education through Share-Net.

O'Donoghue, R. 2002a. *Indigenous Knowledge and Education. Collection of Occasional Papers No. 1*. Grahamstown: Rhodes University Environmental Education Unit.

O'Donoghue, R. 2002b. *Indigenous Knowledge: Towards learning materials and methodologies that respond to social processes of marginalisation and appropriation in eastern southern Africa*. Occasional Paper, Rhodes University Environmental Education Unit, Grahamstown.

O'Donoghue, R. & Janse van Rensburg, E. 1995. *Environment and Methods*. Howick, South Africa: Share-Net.

O'Donoghue, R. & Janse van Rensburg, E. 1999. Indigenous myth, story and knowledge in/as environmental education processes. *Indigenous Knowledge in/as Environmental Education Processes. Environmental Education Association of Southern Africa Monograph Number 3*. Howick, South Africa: Share-Net.

O'Riordan, T. 1998. Sustainability for survival in South Africa. *Global Environmental Change*, 8(2):99–109.

O'Riordan, T. & Cameron, J. 1994. *Interpreting the Precautionary Principle*. London: Earthscan.

Odora, C. 2000. Scenarios for the future of education in Africa: Factors, rationales and starting points for intervention. In: Alexander, N. (ed.). *Educational Innovation in Post-Colonial Africa. Selected Papers from the Pan-African Colloquium*. Cape Town: PRAESA.

Ophuls, W. 1973. Leviathan or oblivian? In: Daly, H.E. (ed.). *Toward a Steady State Economy*. San Francisco: Freeman.

Ophuls, W. 1977. *Ecology and the Politics of Scarcity. A Prologue to a Political Theory of the Steady State*. San Francisco: Freeman.

Page, S. & Page, E. 1991. Western hegemony over African agriculture in Southern Rhodesia and its continuing threat to food security in independent Zimbabwe. *Agriculture and Human Values*, Fall: 3–18.

Palmer, J. 1998. *Environmental Education in the 21st Century*. London: Routledge.

Parker, B. 1997. Understanding the philosophical foundations of outcomes-based education. In: Goolam, F. & Khumalo, L. (eds). *Perspectives on Outcomes-Based Education. Proceedings of a National Conference on Outcomes-Based Education. University of Durban-Westville, 20 March*. Durban.

Patterson, G. 1998. *Dying to be Free. The Canned Lion Scandal and the Case for Ending Trophy Hunting in Africa*. London: Viking.

Peterson, G.E. 1997. *Decentralization in Latin America: Learning through Experience*. Washington, DC: World Bank.

Pezzoli, K. 1997. Sustainable development: A transdisciplinary overview of the literature. *Journal of Environmental Planning and Management*, 40(5):549–574.

Phillips, D.L. 2001. Decentralize Indonesia without dismantling it. *International Herald Tribune*, 23 January.

Popkewitz, T. 1999. A social epistemology of educational research. In: Popkewitz, T. & Fendler, L. (eds). *Critical Theories in Education: Changing Terrains of Knowledge and Politics*. New York: Routledge.

Potter, J. & Wetherel, M. 1987. *Discourse and Social Psychology: Beyond Attitudes and Behaviour*. London: Sage.

Pringle, J. 1983. *The Conservationists and the Killers: The Story of Game Protection and the Wildlife Society of South Africa*. Cape Town: TV Bulpin and Books of Africa.

Quarrie, J. (ed.). 1992. *Earth Summit '92* . London: Regency Press Cooperation.

Quinlan, T. 1995. Grassland degradation and livestock rearing in Lesotho. *Journal of Southern African Studies*, 21(13):491–507.

Rahnema, M. 1997. *The Post Development Reader*. Cape Town: David Philip.

Reijntjes, C.B.H. & Waters-Bayer, A. 1992. *Farming for the Future: An Introduction to Low-External-Input and Sustainable Agriculture*. London: MacMillan.

Rhodes University Environmental Education Unit (RUEEU). 1997. Gold Fields Participatory Course Materials (Theme 3). Grahamstown: RUEEU.

Robottom, I. 1996. Permanently peripheral? Opportunities and constraints in Australian environmental education. *Southern African Journal of Environmental Education*, 16:44–56.

Rohdewohld, R. 2000. Capacity-building needs assessment for local governments in Indonesia. www.capacity.org European Centre for Development Policy Management, Maastricht.

Roth, C. 1992. *Environmental Literacy: Its Roots, Evolution and Directions in the 1990s*. Columbus: ERIC/CSMEE Publications.

Russo, V. 2001. A review of the media in Angola with a view to informing environmental learning and action for change. Unpublished research report.

Sachs, W. 1999. *Planet Dialectics: Explorations in Environment and Developments*. London: Zed Books.

SANParks. 1995. *Corporate Plan. A Framework for Action and Transformation*. Pretoria: SANParks.

SANParks. 1998. *Land Claims Policy*. Pretoria: SANParks.

SANParks. 2000a. *Visions of Change. Social Ecology and South African National Parks*. Johannesburg: SANParks & Development Communications Corporation.

SANParks. 2000b. *Socio-Economic Surveys around Addo Elephant National Park*. Pretoria: SANParks.

SANParks. 2001. *Social Ecology*. Pretoria: SANParks.

SANParks, Danced, GTZ. 2000. *'Towards Best Practice' Social Ecology Conference, 16–18 May, Berg-en-Dal, Kruger National Park*. Pretoria: SANParks/Danced/GTZ.

Sauve, L. 1999. Environmental education between modernity and postmodernity: searching for an integrating educational framework. *Canadian Journal of Environmental Education*, 4, Summer:9–35.

Schacter, M. 2000. Capacity building: a new way of doing business for development assistance organizations. *Policy Brief No. 6.* Institute on Governance, Ottawa.

School Environmental Policy Pack. 1998. Howick, South Africa: Share-Net.

Sellars, R.W. 1997. *Preserving Nature in the National Parks.* London & New Haven: Yale University Press.

Shackleton, S. & Campbell, B. 2001. *Devolution in Natural Resource Management: Institutional Arrangements and Power Shifts. A synthesis of case studies from southern Africa.* Unpublished Special Study, USAID, SADC, NRM Project.

Shiva, V. 1988. *Staying Alive. Women, Ecology and Development.* London: Zed Books.

Shiva, V. 1993. Technology's mythic productions. In: Hallowes, D. (ed.). *Hidden Faces. Environment, Development, Justice: South Africa and the Global Context.* Scottsville, South Africa: Earthlife Africa.

Shiva, V. 2002. *WSSD: World Summit on Sustainable Development or World Summit for Supporting Destruction.* Electronic version available from e-praxis@subdimension.com.

Simasiku, P.A. 2002. *Community-Based Natural Resources Management Principles, Objectives as a Response to Environmental Crisis.* WWF Zambia Education Project (ZEP) Training of Trainers Course in Environmental Education. Kabwe, Zambia: WWF ZEP.

Sparks, J.D. 2002. Beyond oil: transforming global energy. *Newsweek*, 8–15 April: 42–43.

Stenhouse, L. 1975. *An Introduction to Curriculum Research and Development.* London: Heinemann.

Stuart, J., Morojele, M. & Lefoka, P. 1997. Improving our practice. Collaborative classroom action research in Lesotho. In: Crossley, M. & Vulliamy, G. (eds). *Qualitative Educational Research in Developing Countries: Current Perspectives.* New York: Ganand Publishing.

Sundaram, K.V. 1999. Decentralized planning and financing of rural development in India. *Regional Development Dialogue*, 20(2):21–39.

Taylor, J. 1997. *Share-Net: A Case Study of Environmental Education Resource Material Development in a Risk Society.* PhD thesis, Rhodes University, Grahamstown, South Africa.

Taylor, N. & Vinjevold, P. 1999. *Getting Learning Right: Report on the President's Education Initiative Research Project.* Johannesburg: Joint Education Trust (JET).

Thloloe, J. 1997. A new paradigm for journalism in South Africa. *Rhodes Journalism Review*, Number 17. Department of Journalism and Media Studies, Rhodes University, Grahamstown.

Thompson, J.B. 1990. *Ideology and Modern Culture. Critical Social Theory in the Era of Mass Communication.* Cambridge: Polity Press.

Thompson, M. 1998. Environmental Education as an Integral Part of Environmental Management. *Wildlife and Environmental Conservation Society of Zambia Newsletter, July-September.* Lusaka.

Tilley, P. 1995. (ed.). *Our Efforts, Our Rewards: Administrative Management Design for Game Management Areas.* Unpublished report, Ministry of Tourism National Parks and Wildlife Service, ADMADE programme, Lusaka, Zambia.

Tolba, M. 1992. *Saving Our Planet: Challenges and Hopes.* New York: Chapman and Hall.

Tripp, H. 1990. Socially Critical Action Research. *Theory into Practice,* xxix(3):158–166.

Tyler, R. 1949. *Basic Principles of Curriculum and Instruction.* Chicago: University of Chicago Press.

United Nations Conference on Environment and Development (UNCED). 1992. *Agenda 21.* Rio de Janeiro: UNCED.

United Nations Development Programme (UNDP). 1993. *A strategy for assessing UNDP effectiveness in capacity building during the fifth programme cycle.* Central Evaluation Office, Bureau for Policy and Programme Evaluation. New York: UNDP.

United Nations Development Programme (UNDP). 1998. *Zambia Human Development Report.* December.

United Nations Environment Programme (UNEP). 2002. *10 Years after Rio: the UNEP Assessment. Extended Executive Summary.* www.unep.org.

United Nations Educational, Science and Cultural Organisation (UNESCO). 2001. *Teaching and Learning for a Sustainable Future.* CD Rom. UNESCO, Paris.

UNESCO-UNEP. 1978. Tbilisi principles of environmental education. *Connect*, 3(1):1.

VEDCO. 1999. *Document for Sustainable Agriculture and Food Security Project.* Kampala: VEDCO.

VEDCO. 2000. *A Report of VEDCO's Strategic Planning Workshop*, 6–11 August. Kampala: VEDCO.

Villa-Vicencio, C. 1994. Doing theology in context. South African perspective. In: Villa-Vicencio, C. (ed.). *Liberation Theology*. Cape Town: David Philip.

Wagiet, F. 1996. *Teaching the principles of ecology in the urban environment: An investigation into the development of resource materials.* Unpublished Masters thesis. Grahamstown: Rhodes University.

Walker, M. 1990. Action research in South African schools: Gilding gutter education or transformation teaching? *Perspectives in Education*, 11(2):57–64.

Winter, R. 1989. *Learning from Experience: Principles and Practice in Action Research.* London: Falmer Press.

World Bank. 1989. *Sub-Saharan Africa: From Crisis to Sustainable Development: A Long-term Prospective Study.* Washington, D.C.: World Bank.

World Bank. 2000. *Poverty Reduction Strategy Papers: Progress in Implementation.* Washington, D.C.: World Bank.

World Commission on Environment & Development (WCED). 1987. *Our Common Future.* Oxford: Oxford University Press.

World Wide Fund for Nature Zambia Education Project (WWF ZEP). 1995–2000. *Project Document.* Lusaka, Zambia: WWF ZEP.

World Wide Fund for Nature Zambia Education Project (WWF ZEP). 1999. *Environmental Education Manual For Teacher Educators.* Lusaka, Zambia: WWF ZEP.

World Wide Fund for Nature Zambia Education Project (WWF ZEP). 1999. *Fish Conservation Book.* Lusaka, Zambia: WWF ZEP.

World Wide Fund for Nature (WWF). 2000. *The Living Planet Report.* Gland, Switzerland: WWF/UNEP.

Zerner, C. (ed.). 2000. *People, Plants, & Justice: The Politics of Nature Conservation.* New York: Columbia University Press.

Zuber-Skerritt, O. 1996. Introduction: New directions in action research. In: Zuber-Skerritt, O. (ed.). *New Directions in Action Research.* London: Falmer Press.